# More praise for *Opium*

"This detailed and highly readable history of opioid use—and failed policies to contain it—demonstrates convincingly that the best way to address today's epidemic is to acknowledge addiction as the brain disease that it is and treat those who suffer from addiction with the same care and compassion we give to people with other chronic conditions. The recommendations in this book should be seriously considered by anyone concerned with today's opioid epidemic."

> —Congressman Patrick J. Kennedy, member of the President's Commission on Combatting Drug Addiction and the Opioid Crisis

"Highly informed and wonderfully entertaining."

> —Ethan A. Nadelmann, founder of the Drug Policy Alliance

"Halpern and Blistein expertly weave together the many strands of opium's history, from the poppy growers of Neolithic times to the politics of today's opiate epidemic. By learning the whole story and discovering the many erroneous beliefs and misguided policies that have occurred along the way—the reader emerges with a far clearer picture of the problem and what perhaps we can do about it now."

> —Harrison G. Pope Jr., MD, professor of Psychiatry, Harvard Medical School

"Thank God (or whatever higher power you desire) that Halpern and Blistein have done the historical work to demystify the use of opioids. Their research now allows us to focus on the issues that really matter, like keeping users safe and ensuring that patients have access to these effective medications."

—Carl L. Hart, PhD, professor of Psychology, Columbia University and author of *High Price*

"In this lively, irreverent history we learn what Aristotle and William Burroughs, Helen of Troy and Billie Holiday, El Chapo and Thomas Jefferson had in common. They all either used or prescribed, cultivated or profited from opium. The authors chronicle the quackery the drug has inspired, the colonial wars it caused, and the official follies that led to today's opioid crisis—and they outline a fresh and sensible approach to ending it."

—Geoffrey C. Ward, *New York Times* bestselling author of *A First-Class Temperament: The Emergence of Franklin Roosevelt*

"This book takes the reader on a deep journey through the history of opium and how it has shaped medicine, culture, trade, and politics....Halpern and Blistein give readers hope that new policies and treatments to alleviate addiction could make a real difference, if politicians and healthcare institutions are willing to set aside failed strategies that, unfortunately, remain in place."

—Torsten Passie, MD, Goethe-University's Institute for History and Ethics in Medicine

# Opium

# Opium

## How an Ancient Flower
## Shaped and Poisoned Our World

## JOHN H. HALPERN, MD,
## AND DAVID BLISTEIN

NEW YORK   BOSTON

Hachette Books
Hachette Book Group
1290 Avenue of the Americas
New York, NY 10104
hachettebookgroup.com
twitter.com/hachettebooks

First Edition: August 2019

Hachette Books is a division of Hachette Book Group, Inc.
The Hachette Books name and logo are trademarks of Hachette Book Group, Inc.

The publisher is not responsible for websites (or their content) that are not owned by the publisher.

The Hachette Speakers Bureau provides a wide range of authors for speaking events. To find out more, go to www.hachettespeakersbureau.com or call (866) 376-6591.

Print book interior design by Thomas Louie

Photo Credits: p. 1, top: iStock.com/sadikgulec. p. 1, middle: Nikater/Wikimedia Commons/ https://commons.wikimedia.org.wiki/User:Nikater. p. 1, bottom: CC BY/Wellcome Collection. p. 2, top: Than Saffel/Stone Circle Productions. p. 2, middle: Ino Ioannidou and Lenio Bartzioti/American School of Classical Studies at Athens, Corinth Excavations. p. 2, bottom: Miniature from the Post Mundi Fabricam, French codex, first quarter of the 14th century. p. 3, top: Library of Congress, Prints and Photographs Division [LC-USZ62-92635]. p. 3, bottom: Reproduced with permission from Lancashire Archives, Lancashire County Council. p. 4, top: Than Saffel/Stone Circle Productions. p. 4, middle: Unidentified artist, Chinese School, 19th century. Hongs at Canton, China. Oil on Canvas, © 2019 Museum of Fine Arts, Boston. P. 4, bottom: Lithograph after W. S. Sherwill, c. 1850. CC BY/Wellcome Collection. p. 5, top: ©David Blistein. p. 5, middle: William L. Clements Library of American History, University of Michigan. p. 5, bottom: Science Museum, London. CC BY/Wellcome Collection. p. 6, top: Public domain. p. 6, middle: Library of Congress, Prints and Photographs Division [LC-DIG-npcc-28561]. p. 6, bottom: Library of Congress, Prints and Photographs Division [LC-USZ62-87322]. p. 7, top: Library of Congress, William P. Gottlieb Collection, [LC-GLB23-0425 DLC]. p. 7, bottom: Bureau of Prisons Photograph. p. 8, top: U.S. Marine Corps photo by Cpl. Andrew J. Good; the appearance of U.S. Department of Defense (DoD) visual information does not imply or constitute DoD endorsement. p. 8, middle: Ollie Atkins/ National Archives. p. 8, bottom: Courtesy of Vancouver Coastal Health.

Library of Congress Cataloging-in-Publication Data has been applied for.

ISBNs: 978-0-316-41766-2 (hardcover), 978-0-316-41765-5 (ebook)

Printed in the United States of America

LSC-C

10 9 8 7 6 5 4 3 2 1

*John H. Halpern, MD:*
*For Abraham L. Halpern, MD, my father, mentor, and*
*champion for human rights and ethical medical care.*


*David Blistein:*
*For Wendy . . . and the sky that we look upon.*

The problem of chronic opium intoxication...is so extremely complex and far-reaching, so intimately interwoven with public health, commerce, and trade, and social customs, and has evolved so insidiously that we may well ask if the use of opium ever was confined to its sole valuable function namely, that of a therapeutic agent...among the western nations, the United States seems to have acquired the reputation...of being more widely and harmfully affected than any other.

   —New York Bureau of Social Hygiene, Inc., 1928


If there is a war on drugs, then many of our family members are the enemy. And I don't know how you wage war on your own family.

   —Robert Wakefield, *Traffic* (2000)

# Contents

# Opium

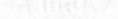

# Preface

---

"How could it be?"

The family and friends of more than 70,000 people say words to that effect every year after hearing that someone they knew died from an overdose.

I never expected to be one of them.

After all, I'm a psychiatrist with a focus on addiction medicine. I know the symptoms and have been trained to spot the warning signs, skills I've honed—or thought I had honed—through decades of practice.

Paul Roderick, as I will call him in an effort to preserve his family's privacy, was my beloved friend—smart, grandly inquisitive, loyal, funny, strong, and an excellent chess player. When we first met, I sensed an undercurrent of childhood rejections, self-parenting, and toxic events in his past, but over the years I watched as he seemed to find the independence and stability he yearned for.

In 1995, I invited him to New Mexico to help me on a federally funded project in which I was comparing the neurocognitive function and mental health of a few disperse groups: Members of the Native American Church (NAC) who had participated in at least 100 prayer services (in which mescaline-containing peyote is ingested for non-drug sacramental purposes) but never had problems with drugs or alcohol; former alcoholics who reported heavy drinking for at least ten years, but had been sober for three months or more; and Navajo tribespeople who never drank or used drugs nor were members of the NAC and didn't regularly participate in their prayer services.

Paul wasn't a researcher at the time. Actually, he didn't know quite *what* he was then or would be. He was working at the mental health center where we met and doing some freelance Spanish translations for a textbook company, while trying to figure out what to do with the rest of his life. One possibility was working in the fields of mental health or neuroscience, and the project seemed a perfect opportunity for him to find out more about both. The fact that he knew Spanish, and could get along with anybody in any language, made him the perfect person to greet our "test subjects" as well as explain the process to relatives and entertain any kids who tagged along so that they did not interrupt their parents' treatment.

I later learned that there was another reason Paul had agreed to join me—a woman named Jenny Anderson, who was the project's whip-smart neuropsychologist. Based solely on my description of her talents, Paul had decided he was in love with her. Paul was like that.

To his dismay, however, he learned from our other team members after arriving in Albuquerque—having traveled 2,000 miles in his beat-up VW Jetta—that Jenny already had a boyfriend. And not just any boyfriend: a young, good-looking, bearded guy named Jeff with a serious trust fund who lived just an hour away in Santa Fe.

To help Paul get over his apparent heartbreak, we drove out to Chaco Canyon, a remote but famous sacred site for the Hopi and Navajo, where there are some of the most remarkable petroglyphs and other pre-Columbian relics in the United States. A pilgrimage to Chaco is a rite of passage for any serious student of Southwest Native American culture.

There was no moon that night, so Paul howled at the Milky Way instead—in despair over the fact the universe was denying him the girl of his dreams, albeit one he had hardly met.

But I persuaded him to stay, and after work each day, Jenny, Paul, and I usually had dinner together, during which he tried to win her over by regularly belittling her boyfriend Jeff in ways that made the fact that she was attracted to him seem utterly absurd.

I should have sensed that these two brilliant researchers were on a collision course. In addition to his irresistibly winning personality, Paul had a number of extraordinary traits going for him—in particular that he looked like Brad Pitt, was in great shape, and loved the outdoors. Perhaps I should have realized it was inevitable that they would get together when she admitted one evening that she had left Harvard for New Mexico because she wanted to do her fieldwork in a place where she could wear a tattoo and drive a pickup.

Shortly thereafter, Jeff was history, and by the time our project was over, Paul and Jenny were living together in New Mexico. While she continued working in neuropsychology, Paul got a job with the University of New Mexico tabulating trends in domestic violence and participating in Native American healthcare initiatives. To my surprise, at thirty-three, he decided it was time to really settle down, so he went to UNM's Medical School. His first job was as a primary-care family physician working for an area Pueblo.

While we rarely saw each other over the next two decades, we talked regularly, sharing stories of our professional lives, marriages, and parenting, as well as our hopes and dreams for the future—one of which was to spend a week camping together in Navajo country with our then ten-year-old sons.

The perfect opportunity presented itself in the late spring of 2017 when we heard that one of our closest friends on the reservation, an

Arizona state judge, was going to officiate his own eldest daughter's wedding. The ceremony would be held in a spectacular sacred setting: a place where the roof was the sky itself; ponderosa pines dwarfed the highest cathedral spires; and rough-hewn logs served as pews. Our plan was to meet up before the wedding and spend a few days camping at an obscure Navajo summer refuge in the Chuska Mountains. It was one of my favorite places in the Southwest, a mini Switzerland where crystal-clear brooks cut through lush high valleys of meadow grass, where the Navajo would take their cattle to cool off from the summer valley heat. Best of all, the area is virtually inaccessible, with sharp, angular volcanic rocks along the rim keeping out anything but the sturdiest off-road vehicles.

I arrived along with my son, Noah, and my fiancée, Ann, before Paul made it there. After surviving the scorching heat in the valley floor, Noah decided to spend the night outside in a hammock with his sleeping bag. Before long, however, the temperature dropped to 20 degrees and he joined us in the back of our rental SUV.

Paul and his family didn't arrive that night. In fact, they didn't arrive the next morning either. Or the day after that.

I tried calling again and again, but just getting a signal on my cell phone was a challenge. Each time, I had to climb onto a mountain ledge, set the phone on a rock, put it on speaker; and as soon as I managed to grab a bar of signal strength, I'd find myself leaving yet another voicemail message:

"Hey Paul! Where are you? I left you the directions. You said you were coming, man. Everyone will be asking me about you, soon, because YOU told ME that you were coming! I mean really...if you aren't going to come then just say so."

I was annoyed, but I was also worried. He lived five or six hours away and weather could throw a wrench in the best plans, but we'd flown all the way out from the East Coast, done that same drive, and arrived in plenty of time. Where was he?

After we came off the mountain and had a reliable signal, I called him repeatedly until he finally picked up. He explained that he and his wife were having some difficulties and one thing led to another and it just couldn't work out. I told him I understood. (After all, I was divorced and traveling with my fiancée.) But actually, I wasn't being honest with him. I *didn't* understand. Relationship problems were one thing, but not showing up for that wedding caused me and our friend the judge unnecessary concern and, more important, showed a disregard for the Native culture we'd both revered all these years.

Paul eventually suggested we come and spend the night before we took our early morning flight back east. We didn't reach his house until close to midnight. Paul's wife and kids were asleep, but we stayed up, talking quietly, long into the night. It felt like old times again.

He apologized for not meeting up with us to camp, especially for the fact that our sons had missed the chance to get to know each other. We talked about coming out again for another camping trip. Mostly, though, we talked about our lives. I was curious about what it was like being a family-practice doctor working for a Native American tribe, and he had lots of questions for me about what it was like to run a for-profit hospital for treating drug addiction—a turn in my career that surprised him as much as it had myself, since neither of us thought I'd ever give up my professorship at Harvard Medical School and practice at McLean Hospital. He said he was proud of me for taking the risk and making the move.

As the conversation went on, I began to look at him with more of a clinical eye. His eyes looked tired, perhaps a little sad, and I was surprised at how much weight he had put on. He never really satisfied my concern about his no-show at the wedding, but I didn't want to press him.

At 4 a.m. we needed to drive to the airport so, after a final hug, we left. I later learned that when he said goodbye to my fiancée, he told her, "I'm sorry. I feel I never got to know you." She took it to mean that he felt bad for not attending the wedding and replied encouragingly that we had all agreed to go camping together another time.

Catching himself, Paul reassured her: "That's right. Of course."

But I never saw my friend again. By the end of the week, he had put a bullet through his residency diploma and another through his head.

His suicide note explained that he had struggled in secret with a drug addiction. Like many addicts, he'd originally been prescribed opioids to deal with a bad knee from running and later chronic back pain. When he tried to stop, he found he couldn't. I would come to learn that he'd tried Suboxone® to help him get through withdrawal, though for some reason he'd stopped taking it.

I couldn't believe he didn't trust me enough to tell me—even though he would have known that I, of all people, would have not just totally understood his pain but had the resources to help him.

Ironically, Paul's addiction was not as bad as that of many of my patients who have recovered. Unlike those patients, however, my friend didn't share his pain, didn't seek out psychotherapy or join a 12-step program for support. He didn't get on methadone or a regulated Suboxone program. He never checked himself into a detox unit or psychiatric facility. And most painful for me, he didn't pick up the phone, even once, to tell his friend who loved him that he needed my help.

\* \* \*

It can be the simple things that hurt most when we lose someone close to us: every chess table I walk by reminds me how we'd promised to spend our retirement playing every chance we got. My heart will forever ache for him.

Now, we are all at a loss: Paul's family, friends, his patients, and all those whose lives he'd made better.

As a physician on the front lines, I try, one patient at a time, to change this dynamic of addiction and death that has somehow emerged from fields of bright red poppy flowers. Every time a patient dies—and yes, *everyone* who works in this field has patients who die—it hurts.

As the late Cardinal Bernardin said to me when I graduated from college, "I used to think there is something extra special about being a Catholic priest, but now that I'm dying of cancer and see all the care provided to me and other patients, I recognize that this 'specialness' can be found in many other professions. Just as it is between priest and parishioner so it is between doctor and patient. We share a moral covenant before God."

Paul's story is an ancient one, and in a strange way it contains for me a message of hope. We are at a cusp of history in which we can develop a more informed approach to opioids. We can understand how drugs derived from (or inspired by) the milky seedpod of the innocent-looking opium poppy have had such a profound impact on the human condition. We can take advantage of new insights and recovery tools to bring about a revolution in the care of opiate abuse and dependence.

Soon, we may have solutions for severe and chronic pain that will keep people from going down the road on which Paul lost his way, as well as remedies that will make it easier to help others find their way back. Until then, the work we must do—to confront the stigma, shame, moral judgments, and self-serving political arguments that swirl around the disease of addiction, as well as create systems that make it possible to help all those who suffer—is enormous and important.

The story of my friend—whose embarrassment became apparent only in retrospect, and whose fear of being stigmatized could have contributed to his decision to end his life—is one reason this book needed to be written. Every year, there are tens of thousands more.

—*John H. Halpern, MD*

# Introduction

Opium is reluctant to give up its secrets.

The physician's painkiller is the addict's poison. The poet's dream is the parent's nightmare. The vigilante's laws are the dealer's opportunity.

The very word *opium* means more things to more people than any other natural substance on earth. In the early 1900s, a famous doctor referred to it as "God's Own Medicine,"[1] while a fiery anti-opium crusader called it "the deadliest foe that has ever menaced [humanity's] future."[2]

Opium wasn't always considered the scourge it is today. Rather, for thousands of years, it was a key ingredient in many formulas developed to relieve pain. Addiction was rare. Overdoses were almost unheard of.

Today, however, opioid addiction is fast becoming the most deadly crisis in American history. In 2017, 47,600 people died of opioid-related overdoses—more than gunshots and car crashes combined...and almost as many as were killed in the entire Vietnam War. The disease

---

1   William Osler (1849–1919)—one of the founders of Johns Hopkins Hospital.
2   Richmond Hobson (1870–1937), "The Struggle of Mankind Against Its Deadliest Foe." Also, see Ch. 26.

is straining our prison system, dividing families, and defying virtually every legislative solution to treat it.

Opium's history is as complex as it is disturbing. To understand it, we need to examine the papyruses of ancient Egyptian scribes for whom opium was an essential ingredient in a wide range of remedies; the edicts of Chinese emperors who watched in horror as addiction debilitated large swaths of the population; the propaganda of anti-drug vigilantes whose scare tactics often only increased the awareness and use of opioids; and, finally, the latest breakthroughs in prevention and treatment that hold out the promise of real solutions.

We need to watch Portuguese sailors blend opium with tobacco to make a more potent smoke; a nineteenth-century Chinese official destroy 3 million pounds of British opium in a futile attempt to end *his* country's opioid epidemic; and, at the end of the twentieth century, a modern pharmaceutical company develop a medicine for end-of-life pain that has turned into the deadliest drug in history.

In particular, we must learn from our past attempts to solve the problem—and how some of those attempts have only made the problem much, much worse.

# PART I

## Opium in Antiquity

# Chapter 1

# The Mysterious Origins of the Opium Poppy

Though we can learn many things about the origins of opium by studying archaeology, ethnobotany, and genetics, to understand the role it played in ancient societies, a little mythology comes in handy.

According to no less an authority than William Burroughs, the notorious heroin addict, cultural icon, pioneer of drug use among the "Beat Generation," and author of the controversial novel *Naked Lunch*, the first opium addicts were members of a Cro-Magnon tribe called the Unglings who lived in the Alps 30,000 years ago.

In a marvelous and presumably drug-infused fantasy called *God's Own Medicine*,[1] Burroughs imagines members of this tribe as "Homo sapiens like you or me—or the folks next door, wearing animal skins and carrying stone axes." He goes on to propose that it's late summer and they are walking through a field of flowers. One of the elders, crippled by rheumatism, falls to his knees. He finds himself face-to-face with a strange-looking seedpod oozing a crusty brown sap. It smells really bad, but looks sweet like honey, so he tastes some. Sadly,

---

1   The legend was originally published by Burroughs in his book *God's Own Medicine*. It also served as the introduction to Latimer and Goldberg's classic *Flowers in the Blood*, the first book to present contemporary "counter-cultural" drug issues in the light of opium's history. We've paraphrased and embroidered Burroughs's story a bit...but not much.

it's terribly bitter. Feeling worse than before, he drags himself to his feet and stumbles forward.

A while later, he feels less pain in his joints, stands up straighter, and begins to walk without staggering. A few steps farther along, he throws his staff to the ground, looks to the heavens, and starts thinking thoughts that he never thought before—that perhaps no human brain has ever thought before.

One of those thoughts is that he should tell his people to gather as many of those pods as they can, scrape off the dried sticky stuff, and fill as many gourds as possible for future use.

That winter is far more pleasant than usual.

By the next spring, however, Burroughs imagines that the sap is all gone and the old Ungling's people appear bereft. They're not in the mood to do any hunting or gathering or fishing. As his joints begin to creak again, the wise elder, who is now perceived to have magical powers, has a vision (or so he says) that they need to go back to where they camped that fateful day the year before in order to collect more of those pods. For the first time, the movement of a people is driven by a hunger for a drug rather than food.

Allegorical as the story may be, it is this dependency—which Burroughs understood all too well—that would eventually make opium a versatile and valuable commodity in world trade, resulting in strange wars and even stranger bedfellows.

The empirical evidence for opium's origins isn't all that much more reliable than Burroughs's legend. Researchers have spent their entire careers laboring at remote archaeological sites trying to find it. They've examined inscriptions on pottery and pictographs on papyruses. Some have looked through undisturbed firepits and pre-historic privies for poppy pods or even individual seeds, or subjected skeletal human remains (e.g., tissue, hair, teeth, bone marrow) to radioimmunoassay, gas/liquid chromatography, and mass spectro-metry.[2]

Unfortunately, finding and testing samples of opium residue in or on an artifact that's been exposed to thousands of years of wear and tear—not to mention possible contamination by random acts of unkind vandalism—is a whole lot harder than testing a batch of street heroin to see if it's pure. As a result, we can't be sure whether opium use originated in southwest Europe, southeast Europe, central Europe, the northern Mediterranean, or Asia Minor—or even Egypt, northern Europe, Great Britain, or various Asian locations.

The best way to find the origin of a plant is to find one growing in the wild with similar DNA, which would indicate how an existing wild plant had mutated into a new cultivar (either by accident or human breeding). However, no one has ever found a legitimate ancient wild ancestor of the opium poppy, *Papaver somniferum*.

While researchers *have* discovered other varieties of poppies that make some of the same alkaloids as the *opium* poppy, this single species of a single genus of a single family is the only one with any significant amount of morphine.[3]

The first evidence that Neolithic hominids and the opium poppy crossed paths was discovered in 1854 in the small Swiss town of Mellen on the southeast shore of Lake Zurich.[4] That winter was

---

2   Veiga, "Opium: Was it Used as a Recreational Drug in Ancient Egypt?"
3   Scientists define alkaloids in different ways. Typically, they are chemicals found in plants that can affect human and animal metabolism—e.g., slow down hearts or dilate veins. Some, like strychnine, also make great poisons. There are two dozen different alkaloids in the opium poppy. Only three are narcotic—morphine, codeine, and thebaine. Scientists are in the process of figuring out which genes are responsible for the opium poppy's ability to make morphine...which would enable them to insert those genes into other plants. See Reuters, "Scientists Find Gene Secret."
4   Mornbelli, "Lake Dwellings Reveal Hidden Past."

historically dry and cold—bad for kids who went off to school without their mittens, but good, it would turn out, for local entrepreneurs who wanted to turn their town into a tourist destination for folks from Zurich.

Until then, those tourists had just been waving at the locals from steamers that took them on leisurely summer cruises down the lake, because the city didn't have a place for those boats to dock and, therefore, was missing out on all the Swiss francs the tourists might spend eating, drinking, and buying local crafts. But the drought that winter exposed the shores of the lake bed, giving the people of Mellen the opportunity to set log footings into the sand without working underwater. Imagine their surprise when they started digging and discovered that someone had been there first: buried in the lake bed were dozens of poles nearly 7,000 years old.

Mellen's prehistoric inhabitants had not been building docks for tourists from Zurich, but rather constructing pile dwellings—long communal buildings built on stilts. The raised structures protected them from flooding, wild animals, and enemies, while giving them easy access to water and arable, unforested land. Plus, in wet seasons, they could fish through the living-room windows. These pole buildings— prototypical communes—were up to 200 feet long. Out back, the lake people grew wheat, apples, peas, lentils, flax, barley, and, as it turned out, poppies, which could have met their nutritional as well as their analgesic needs.[5]

Another early sign of poppy use was discovered in the 1990s at an underwater dig at a site north of Rome called La Marmotta. The

---

5   A single tablespoon of poppy seeds has 1.6 grams of protein, the same as sesame seeds—although slightly less than flax seeds (1.9 grams) and chia (2.0 grams). The grand prize goes, unsurprisingly, to hemp seeds (3.2 grams).

site flooded around 5700 BC, preserving the settlement mostly intact. There, archaeologists found not only remains of what appear to be poppy seeds, but also of a thirty-five-foot-long canoe dug out from a single oak tree, which would be perfect for sailing off into the Mediterranean to develop early trade routes.[6]

Poppy remains have also been found at Neolithic settlements in the Rhine, Rhône, and other river valleys in Europe, where people had discovered not only how to build settlements and to farm, but how to make functional objects. Two of the advanced tribes in these areas are particularly famous for their distinctive styles of pottery. One is known as the *linearbandkeramik* (abbreviated LBK) because of the linear bands that decorate their pots. The other group, La Hoguette, settled in the Rhône Valley, where its people tempered their clay with bone for greater strength and decorated their pots with impressed or incised images.[7] Archaeologists have found poppy residue in *both* types of pottery, suggesting that the groups interacted through trade and/or war.

Halfway across the continent, another culture—this one in southern Spain—was growing poppies as well. In a cave called Cueva de los Murciélagos, poppy capsules were found woven into grass baskets that had been placed next to human skeletons, along with other signs of respect for the dead, such as flowers, sea snails, and gemstones. In the bones of one skeleton and the jaws of another were traces of actual opium consumption.[8] These early Spaniards were clearly using the juice of the opium poppy either to ease the pain of dying or prepare the person spiritually for a pleasant journey to the hereafter— possibly both.

---

6  Kunzig and Tzar, "La Marmotta."
7  Reingruber, Tsirtsoni, and Nedelcheva, *Going West?*, 79ff.
8  Guerra-Doce, "Psychoactive Substances."

It will take a combination of new archeological discoveries and advancements in DNA testing to determine more precisely where poppy use began, and how exactly it made its first moves across the world. What we do know, however, suggests that by approximately 3000 BC the opium poppy had worked its way to Sumeria, where in 1893 a group of archaeologists from the University of Pennsylvania unearthed a treasure trove of clay tablets while excavating the city of Nippur near modern Baghdad.[9] This walled city on the Euphrates River had been home to more than 40,000 inhabitants, most of whom were farmers, shepherds, craftspeople, and scribes. The river flooded seasonally, leaving soil that was easy to shape into tablets for scribes to record everything from details about the latest harvest to legends about the Great Flood. After being left to dry quickly in the blazing sun, these tablets encoded information for millennia.

The most famous contribution to our understanding of ancient use of opium came from Robert Dougherty, a professor at Yale, who spent much of his academic career painstakingly cross-referencing the tiny ideograms scratched into the tablets. On the back of one, he found two symbols that he tentatively *transliterated* (or anglicized) as *hul gil* and *translated* as "joy plant." Though he cautioned in his translation that "Gil as a single ideogram represented a number of plants" and "its meaning in the ideogram for 'opium' is difficult to determine with exactness,"[10] his hesitancy didn't stop writers of every opium-related book, magazine article, podcast, and online forum since to repeat the "joy plant" translation, thereby implying that the Sumerians used opium recreationally, even frivolously.

As one anthropologist put it: "Decoding humanity's early use of the

---

9   Swaminathan, "Exhibitions."
10   Terry et al., *The Opium Problem*, 55.

opium poppy leads us along the tangled paths of ancient languages, trying to discern the meanings of signs as they change over time and from one culture to another."[11]

This is not only an academic discussion. Four thousand years later, simplistic attitudes and cultural differences *continue* to be major obstacles in the way of a constructive conversation between the West and the East about opioids. For example, while Western leaders such as President Trump have accused Eastern labs of poisoning us with opium, people there speak with disdain about those in the West poisoning *themselves* with alcohol.[12]

Whether people in the ancient Middle East were using the poppy for pleasure, insight, or pain relief, there is no indication that they thought doing so was in any way *wrong*. Western governments trying to understand Afghani farmers share a major problem with academic researchers: they truly are not speaking the same language.

By the seventh century BC, opium was most certainly being used for its pain-relieving properties in Assyria. There King Ashurbanipal—who came to power perhaps in a fratricidal coup when his brother died suspiciously and who would eventually conquer Babylonia, Persia, Syria, and briefly Egypt—built a remarkable library in his capital of Nineveh. The 2,000 tablets in this library have many cuneiform symbols that are similar to the Sumerian ones. When a scholar named Reginald Thompson started studying them, he not only saw a lot of *hul gil* ideograms; he also found enough symbols to conclude that

---

11  Saunders, *The Poppy*, 12.
12  Higgins, "Trump Calls on China to Seek Death Penalty for Fentanyl Distributors." As one writer puts it: "Allah forbids alcohol but not opium whilst the West forbids opium but permits alcohol: opium should not be banned internationally as this would cut across a socio-religious, cultural aspect of Islamic life. If, the argument goes, the West wishes to internationally ban opium then it should also globally ban alcohol." Booth, *Opium*, 341.

the Assyrians used opium poppies as well as other narcotics to treat headaches, bruises, eye problems, and all kinds of ills associated with pregnancy and digestion. They even took it as a "juice" or elixir called *arat. pa. pa.* or *aratpa-pal.*[13]

One piece of Thompson's evidence shows just how central the flower's power had become in this ancient culture: there are images of Nisaba, the goddess of fertility, with poppy pods on her shoulders as well as bronze pendants in the shape of poppy seed pods, with incised slits where the pods are traditionally cut to harvest the sap.[14]

The Assyrians not only used opium in medicine, they worshipped the goddess who had blessed them with it.

13  Dormandy, *Opium*, 8.
14  Kritikos and Papadaki, "The History of the Poppy and Opium in Antiquity."

# Chapter 2

## Papyruses and Poppies

At the same time the Assyrians were using the gooey seed extract of the opium poppy for healing and ritual, it was becoming an important ingredient in Egyptian medicine.

Much of what we know about Ancient Egypt consists of carefully inscribed hieroglyphs on deteriorating and torn fragments of a plant called papyrus. One of the first that saw the light of day appeared in the late 1860s when a man from Orlando, Florida, named Edwin Smith moved to Egypt and bought a valuable scroll from an antiques dealer who told him it had spent the previous 3,000 years between the legs of a mummy. It was twenty meters long, 110 pages, and bore 876 prescriptions for everything from high anxiety to crocodile bites. The opium poppy figured prominently in the cures listed in this early pharmacopeia:[1]

- Headache: Take a mortar and pestle and smash together some poppy berries, coriander berries, wormwood, juniper berries, and juniper honey. (This is the medicine the goddess Isis prepared for the shape-shifting Supreme Sun God Ra.)
- Constipation: "Drive out the excrement in the body" by mixing

---

1   Dormandy, *Opium*, Ch. 1.

up some "poppy stalk, castor oil, dates of the male palm, cyperus grass, coriander, and cold beer."[2]

- Constipation compounded by flu-like aches and pains: The healer will proclaim, "Death has penetrated [your] mouth and taken up its abode," and, as a last resort, mix up a "stinging remedy" of tehua berry, poppy, peppermint, annek, and red sexet seeds.

- Opium was also used as an ingredient to fortify a person's constitution—which they referred to as *met*.

In modern times, we often speak of the challenge of balancing opium's palliative and addictive qualities as if they represent a zero-sum game. The notion of opium being part of an everyday health regimen may seem surprising, even incongruous. And yet, that was one of its most important uses back then—in the same way that we might take a vitamin supplement today.

To make the *met* more "supple," for example, the scribes recommended a topical application of Bullock's fat, Yeast-of-Wine, Garlic, Saltpetre-from-the-South, Poppy-berries, and Oil-of-Myrrh.[3]

While at first glance some of their remedies and rituals might seem strange, there may not be much difference between the effect of their complex natural formulas and our obsessively targeted medicines; their priests' suggestions to put encouraging messages to the gods on their bodies and our, at times, inspirational modern-day tattoos; their sacred baths and our health spas; or their healing amulets and the therapeutic crystals people wear on necklaces or place in strategic locations in their homes.

---

2  Bryan, "Ebers Papyrus," 45. It's not clear why the beer had to be cold or what form of "refrigeration" they were using.

3  Ibid., 113ff. (NB: All spellings, including hyphenations, are [*sic*].)

* * *

Among the thousands of prescriptions Egyptian doctors were writing at the time, one, in particular, would prove deadly—the "Remedy to Prevent the Excessive Crying of Children." There were a number of variations, but the two essential ingredients were grains from the *shepenn* (poppy) and fly droppings (or "dirt").[4]

The ingredients were supposed to be pulped, sieved, and then taken for four days, at which point the poor drug-addled kid would allegedly stop crying "at once."

One of the few facts that virtually every Egyptologist agrees on is that children were the first Egyptians to die of opium overdoses. The jugs with poppy paintings in children's sarcophagi are not only hints of what took them prematurely to the spirit world, but also sad harbingers of the fact that, today in America, a baby is born every fifteen minutes suffering from opioid withdrawal.[5]

* * *

It appears that opium was also important in the afterlife. Some Egyptian royalty were buried with jugs that had slightly bent necks that resembled poppy plants, suggesting that the drug might prove

---

4  There are often modern analogs of strange ancient ingredients. For example, the *Musca domestica* (house fly) is used today as a homeopathic remedy for headaches, toothaches, and mumps.

5  National Institute on Drug Abuse, "Dramatic Increases in Maternal Opioid Use and Neonatal Abstinence Syndrome." Also see Vanderbilt University Medical Center, "Neonatal Abstinence Syndrome and Opioid Policy." Even a baby born to a mother in recovery who is taking Suboxone will experience symptoms, as could the child of a mother taking OxyContin for chronic pain. It's important for a pregnant woman to give her healthcare professional this information as early in the pregnancy as possible.

useful for pain relief at the end of one life or beginning of the next. In one tomb a fabulous carnelian necklace was discovered with dozens of poppy-shaped beads.

But in ancient Egypt, where medicine and spirituality made the world go around, while opium was central to both, there was, in fact, no native opium grown. Rather, like today's junkies, the Egyptian priests needed to import the sap, and they would do so at most any price.

# Chapter 3

# A Journey Around the Mediterranean[1]

Traders began crisscrossing the Mediterranean around the four-teenth century BC, slowly building a supply chain for a wide range of goods including opium-inspired artifacts, the discovery of which have proven invaluable for understanding the way trade developed between cultures and the role the drug played in the early days of global trade.[2]

Before casting off, Greek traders would load up with foodstuffs to eat and sell at ports along the way. Their cargo also included finely crafted six-inch-tall ceramic juglets from Crete as well as amphoras about three times as large. The juglets, like those found in Egyptian tombs, had narrow necks and smooth and bulbous bases that looked like upside-down poppy pods. They also featured decorative lines placed exactly where a skilled opium harvester would incise the pods.[3]

---

1   Day, "Botany Meets Archaeology"; Kritikos and Papadaki, "The History of the Poppy and Opium in Antiquity"; Merrilles, "Opium Trade in the Bronze Age Levant"; and Merlin, *On the Trail of the Opium Poppy* are the main sources for this chapter.
2   They traveled on boats that were far sturdier than we might imagine—about 50 feet long and 15 feet wide, constructed from Lebanese cedar and held together by tight mortise and tenon joints.
3   Some researchers still argue that the shape of those amphora is more reminiscent of pomegranates, that the jugs held pomegranate juice, and that it is simply a coincidence that the decorative lines resemble the slits cut into poppies. But references from the time suggest that they are grasping at straws.

The most important thing these traders would do while preparing for their voyage was listen for rumors floating around the agora about when the next shipment of opium would arrive. When it did, they'd try to be first in line. After examining and tasting the product, and haggling a little over the price, they would buy enough to fill their jugs— large and small—in order to make both wholesale and retail sales.

Dealers jealously guarded the source of their product. For all that the Greek traders knew, it could have come from the other side of the world. Actually, the poppies were likely grown just a few hundred miles or so away in modern-day Turkey or Iran.

Casting off, a trader's first stop might be the island of Rhodes, about 450 miles to the east, where buyers would pay for the opium with some of their own drug paraphernalia—in particular long hollow pins with heads in the shape of opium capsules, which suggest that the ancient Greeks also smoked opium.[4]

Hugging the eastern coastline, they might next stop in Syria, where the Greeks could trade their product for elegant necklaces with poppy-shaped beads. The island of Cyprus, 150 miles back out into the Mediterranean, also had a burgeoning trade in poppy-shaped pins and juglets. None of these accoutrements took up much space on the boat, and all would be easy to sell back in Greece where there was a ready market for any interesting new drug paraphernalia.

Finally, the boat would arrive in Egypt where the real money was. There, traders were able to sell opium up and down the Nile as easily and profitably as a street dealer with a new shipment of heroin in twenty-first-century America.

---

4  Portuguese explorers in the 1500s are usually credited with spreading the practice of opium smoking. Artifacts like these suggest people were doing it much earlier.

Traders followed similar routes around and across the Mediterranean for centuries, buying and selling an increasingly wide range of commodities. Opium's influence on trade changed significantly sometime in the middle of the first millennium BC when Egyptians began growing their own supply. Until then, they had been able to produce only a small amount of their own opium from *Papaver rhoeas*, a variety that they had long used ornamentally in their gardens, but that lacked the strong psychoactive properties of the imported Turkish variety *Papaver somniferum*.[5]

But, eventually, the Egyptians obtained seeds of the Turkish variety or figured out how to hybridize their red poppy to yield more potent opium, which they began growing in the rich Nile Valley near Thebes. It yielded such a high-quality opium that it earned a Latin name of its own, opium *Thebaicum*.

Egypt's ability to produce its own opium poppies may not have had a huge effect on the balance of trade in the Mediterranean. Back then agricultural shifts didn't happen in one season. Now they do. As we'll see, when a major drought devastated most of the Afghani poppy crop in the early 2000s, it almost bankrupted the Taliban, but there was little effect on the worldwide market. The source of supply simply shifted, in a growing season or two, east to the Golden Triangle. In the same way, crop eradication in one area of Afghanistan today simply leads to increased production in another area.

---

5   *Papaver rhoeas* is also known as the corn poppy and Flanders poppy because of the profusion of wild seeds that sprouted on the devastated killing fields of World War I. See Saunders, *The Poppy*. *Papaver rhoeas* does have a small amount of morphine and mildly sedating rhoedine in its molecular brew. See Montag, "How to Make Opium from a Papaver Rhoeas." While only minimally narcotic, every part of the red poppy—berries, seeds, grain, and stalk—had a medicinal use in Ancient Egypt.

* * *

Archaeological finds tell us only so much about the prevalence and use of opium in the ancient world. Another way is to study legends and myths, because merchants don't just deal in products, they deal in ideas. Trade routes are cultural routes as well. The strands of truth in travelers' tall tales are eventually embroidered into legend.

*The Odyssey*, for example, reveals how familiar Egyptian and Greek royalty were with opium. After Helen was rescued from Troy, she began to spend a lot of time playing the zither and blending herbs. One of her blends was called *nepenthe*, which has been translated as "anti-sorrow" or "drug of forgetfulness," a formula she claimed to have been given by an Egyptian queen named Polydamna. It was a good drug for Helen to keep in stock because she frequently had to deal with veterans of the Trojan War who were suffering from what we would now think of as severe PTSD.

For dinner one evening, the esteemed guests included her recently returned husband Menelaus and Odysseus's son, Telemachus. As they eat, the men "begin to cry as they remember the comrades whom they lost in in the war." They save their biggest tears for Odysseus, whom Menelaus cannot even think about without losing sleep. In Greece, however, "real men" literally did not cry—it was considered in especially bad form at dinner where it would have been considered an insult to the host. So, Helen "cast[s] a drug into the wine whereof they drank, a drug to lull all pain and anger, and bring forgetfulness of every sorrow," and before you know it, Menelaus is happily telling war stories about the horse again. In fact, the next morning he tells Telemachus about how he learned the truth of his father's death from the sea god Proteus, whom he and his men briefly captured after spending the night hidden under foul-smelling seal skins. The story is so wild that it makes you wonder whether maybe the "nepenthes" hadn't worn off yet.

Opium use was also common in the legends of the Eleusian Mysteries that took place twelve miles from Athens. These rituals were held with the ostensible purpose of honoring Demeter, the goddess of fertility, and her daughter Persephone, and involved music, drumming, incense burning, and drug taking. A key part of the ritual involved a chest known as a *kiste* and basket called a *kalathos*, the latter of which contained a mind-altering drink known as *kykeion*. Only the highest initiates knew what was in it. All the ordinary townspeople knew was they came out of the temple feeling a whole lot different from when they walked in.

There's been a lot of speculation about the formula for *kykeion*. The most popular urban psychedelic legend—which is backed by an increasing amount of scientific evidence—is that it included some form of ergot fungus, a precursor to LSD that would have grown widely on grasses and grain crops there. But that interpretation misses a clue that stared the archeologists who discovered the site where the Mysteries were held right in the face: when they descended into Demeter's temple, like the participants of the Mysteries had millennia before, they passed a giant statue of the goddess herself, wearing a basket-like crown festooned with poppies.

# PART II

## Opium and the Birth of Modern Medicine

# Chapter 4

# Classic Cures, Ancient Addictions

All the major tensions surrounding opioid use today can be found in the manuscripts of ancient Greek and Roman writers. They describe physicians walking the fine line between poison and cure; users risking addiction and overdose; soldiers dealing with pain or desperate for courage; philosophers and poets discovering wisdom and delusion; political scandals, assassinations, and, in one case, a death by suicide. It's a part of history that many classical scholars prefer to ignore, but it is important for us to understand in order to debunk the longtime Western assumption that countries in the Far East are at the root of today's opium problem, in terms of both cultivation and use.

In his book *The Chemical Muse: Drug Use and the Roots of Western Civilization*,[1] the author James Hillman tells the cautionary tale of his Ph.D. dissertation defense. It had taken him ten years and thousands of dollars to get to that moment: appearing before a committee of five professors to defend his 250-page paper on medicinal drugs in Roman times. After several hours of rigorous questioning, four of the five left the room—without a single smile, compliment, or handshake—leaving him alone with his adviser. Things didn't look good. As it

---

1   Hillman, *The Chemical Muse*, 6ff.

turned out, however, the situation was not hopeless. All he had to do was take out the chapter related to *recreational* drug use (and all other references to abuse of narcotics) because, as far as the chairman of the history department was concerned, Romans "just wouldn't do such a thing." But, as Hillman demonstrates in his book, they did. People in ancient times accepted opioid use—with all its potential pain-relieving benefits and mortal dangers—as simply part of the continuum of the human relationship with the natural world, intermediated by a pantheon of gods.

\* \* \*

There were hundreds, probably thousands of people practicing medicine of some sort in ancient Greece and Rome, and while some relied on tried-and-true folk medicines and others were shameless quacks, many made clinical observations of symptoms that we still don't fully understand, including drug tolerance, dependence, addiction, overdose, withdrawal, maintenance doses, and adulteration. The most famous physician of them all was Hippocrates, who practiced his medical arts around 400 BC.

Most doctors today would say that the Hippocratic Oath—"First do no harm"—was his most significant contribution to the history of science, but his insistence that diseases were *not* caused by divine retribution but rather by bodily imbalances is equally significant.

He said there were four fundamental substances in the body called humors—blood, phlegm, yellow bile, and black bile—which were naturally present in different degrees in four different types of people—sanguine, phlegmatic, choleric, and melancholic.

The language may seem unusual, but they established an important precedent—that it's important to understand *through clinical observation* why some people get sick and some don't when exposed to the same environment; why some people get better and some don't when exposed to the same drug; and why the seasons affect patients differently, as does what they eat and drink, and whether or not they exercise.[2]

Hippocrates's categorization of humors and types might not be as finely tuned as our science of genes, and his naked eyes might not have been as perceptive as our MRIs, but he was as keen an observer of human physiology as anyone who has lived before or since. The idea of "addictive personalities" would have made perfect sense to him.

Hippocrates was an equally keen observer of the properties of plants and recognized opium as a valuable hypnotic, narcotic, and cathartic.

For example, to restore balance to the humors, he said it could be spread over the forehead, stuffed up the nose, or used as a suppository.

He also recommended opium for specific ailments, including abdominal and gynecological pain, and he considered it a last resort for people with anal inflammations—perhaps opium's tendency to constipate may have simply given the patient a little rest down there to allow more time for healing.

Hippocrates recognized the psychological applications of opium too, suggesting it for insomnia or even outright mania. In that regard, one of his most interesting applications was for "confronting the consequences of being chaste while palpitating with lust." In this case, he doesn't say exactly how or where to apply it, or what those consequences were, although we can make a good guess.[3]

---

2    Margotta, *The Story of Medicine*, 68.
3    Escohotado, *The General History of Drugs*, 225.

* * *

A century or so after Hippocrates came up with a cure for palpitating with lust, one adolescent, in particular, was learning about opium, presumably just in time.

His name was Alexander the Great, though at the time he was just the thirteen-year-old Macedonian son of King Philip and Queen Olympias. Wanting him to get the best possible education, Philip asked the famous Aristotle to tutor the boy on all kinds of subjects including geometry, astronomy, botany, and medicine.

At the time, Aristotle was living on the island of Lesbos trying to figure out how the world worked by using observation and reason rather than magic and mysticism. To do so, he experimented with many plants, including mind-altering ones like the poppy.

Before Alexander arrived, Aristotle had begun mentoring Theophrastus (371 BC–287 BC), who is now considered the father of botany. In his two-volume *Enquiry into Plants*, Theophrastus provides an exquisite description of the opium poppy, including how to collect "juice" directly from plants (as opposed to soaking them in water or alcohol or boiling).

He also describes its usefulness as a poison—how, when combined with the juice of hemlock (and/or henbane), it led to "an easy and painless end"; plus, it had a "conveniently small size, weighing only somewhat less than a quarter of an ounce," held its strength (or "virtue") over time, and there was no antidote.[4]

As important were his insights into the phenomenon of tolerance, which he describes in terms that are equally true today: "The virtues

---

4  A generation before, Socrates had taken a poison famously referred to as just hemlock. It likely had a few other "top notes," including poppy.

of all drugs," he wrote, "become weaker to those who are accustomed to them, and in some cases become entirely ineffective....It seems that some poisons become poisonous because they are unfamiliar, or perhaps it is a more accurate way of putting it to say that familiarity makes poisons non-poisonous; for, when the constitution has accepted them and prevails over them, they cease to be poisons...besides the constitution, it is plain that use has something to do with it."[5]

Although Alexander only studied with Aristotle and Theophrastus for a few years before Philip made his precocious son a deputy military commander at fifteen, there was plenty of time for the young empire builder to learn these details of the poppy's use and abuse.

In Alexander's culture, opium was largely an upper-class indulgence. To the rich, it offered an escape from the trials and tribulations of wealth, power, and luxury. In Athens, it was particularly fashionable among the city's golden youth.[6] On his campaigns, however, Alexander would have been far more interested in opium's painkilling properties.

In the conquests that made him famous, Alexander and his bellicose friends traveled 5,000 miles across modern-day Turkey, Syria, Iraq, Iran, Afghanistan, Tajikistan, Uzbekistan, Pakistan, India, and back through Lebanon, Jordan, Israel, and Egypt, more or less in that order, sometimes marching as much as forty miles a day, fighting bloody battles at every turn and experiencing no end of pain.[7] During the course of his campaigns, he got whacked in the head so hard it broke his helmet in two (Battle of the Granicus River, 334 BC); had a sword plunged into his thigh (Battle of Issus, 333); was struck by a missile

---

5   All quotes from Theophrastus, *Enquiry into Plants*, IX and XVII.
6   Dormandy, *Opium*, 19.
7   See Engels, *Alexander the Great and the Logistics of the Macedonian Army*, 102.

from a catapult that pierced his shield and wounded him seriously (Gaza, 332); took an arrow through the calf (Tanais River, 329); was struck by a stone that hit his head and neck (Cyropolis, 329); was hit by a dart through the shoulder (Battle with the Aspasians, 327); took an arrow in the ankle, Achilles style (Massaga, 327); and took another arrow through his lungs (Mallian River, 326), which bled badly and knocked him unconscious.[8]

Opium was undoubtedly part of Alexander's traveling medicine cabinet and would have provided some relief from these excruciating injuries as well as any PTSD. Along with the copious amounts of wine that he was famous for serving at banquets, opium likely fueled his megalomaniacal vision—his determination, reckless persistence, and ultimate downfall.[9]

More important is the impact Alexander, one of the most famously flagrant drug abusers in history, would have had on others. He was charismatic, a leader followed by thousands of loyal troops. And, whether he was reviled or beloved in the lands he conquered, his actions and behavior were known far and wide and would have been imitated and emulated. Just as societal changes followed in his army's wake, so would have alcohol and drug use.

* * *

By 300 BC, the Greeks knew of multiple medical uses of opium: its effectiveness in helping soldiers cope with the pain and trauma of war; the phenomenon of tolerance, whereby a drug would become

---

8   Mann, "Alexander's Injuries Part 1." Apostolakis et al., "Alexander the Great's Life-Threatening Thoracic Trauma."
9   See Gabriel, *The Madness of Alexander the Great*. A comprehensive look at Alexander and his armies from the modern perspective of PTSD.

less effective over time; and the fact that one person's medicine could be another's poison—a conundrum at the heart of many of our perceptions and misperceptions of narcotic use today.

As Socrates discovered when forced to drink hemlock, the ancients also knew how to use opium and other drugs in assassinations, executions, and suicides. At a time when 13,000 Americans are murdered with guns every year and another 22,000 use guns to commit suicide,[10] it is easy to forget that for most of human history people have used drugs to kill each other and end their own lives.

In fact, the ancients—particularly rulers and others in power—had much the same fascination with poisons and their antidotes as we do with firearms today. The most well-known was Mithridates VI (135–63 BC), the king of Pontus in the eastern part of Turkey. He is famous for his attempts to develop a *mithridatum*, which has become a synonym for "antidote."

His contemporary, a king named Attalus III (170 BC–133 BC), who ruled Pergamon (now part of western Turkey) turned the search for herbal poisons into an obsession.[11] One of his hobbies was mixing "noxious herbs with harmless ones" in order to find the most painful way possible to execute his enemies, while using the same knowledge to find a universal antidote (*theriac*) in case anyone tried to do the same to him.

As part of his research, Attalus performed one of the first controlled clinical trials in history, ordering various potions—some presumably benign, others possibly deadly—to be given to criminals in order to see how well those poisons (and antidotes) worked.

---

10   CDC data for 2017.
11   Reddit, "Why Did King Attalus III of Pergamon Give His Country to the Roman Republic in His Final Will?"

Attalus's research assistant was an herbalist named Nicander of Colophon, who had Homeric aspirations.[12] He said that if you drink too much opium you'll fall asleep, *completely* asleep. First, your extremities will be "chilled; eyes do not open but are bound motionless by their eyelids. With the exhaustion an odorous sweat bathes all the body, turns cheeks pale, and causes the lips to swell; the bonds of the jaw are relaxed, and through the throat the labored breath passes faint and chill. And often either the livid nail or wrinkled nostril is a harbinger of death; sometimes too the sunken eyes."[13] As an antidote, Nicander recommended hot wine and a syrup made from grapes, oil of roses and iris, and olive oil. If that didn't work, he suggested slapping the person around to try to get him or her to vomit, thereby "ridding himself of the grievous affliction."[14]

Nicander also did an exhaustive study of snake and spider venoms, which were a threat equal or greater to that of opium to rulers and ruled alike. Hillman's theory is that Nicander's exhaustive research was really inspired by his search for the hallucinogenic sweet spot that, in some cases, can be found in the smooth continuum between medicine and poison. He suggests that his most famous work, *Alexipharmaca*, is more about the proper use of mind-altering substances and treating overdoses than committing murder.[15]

Assisted suicide is yet another modern practice involving poison that has roots in ancient Greece. The famous historian Pliny the Elder said that the juice of the black poppy "is productive of sleep until death" and described it being used by the terminally ill.[16]

---

12   He wrote his observations in as poetic a manner as he could get away with in a
      scientific text. Grout, "Nicander."
13   Nicander of Colophon, *Poems and Poetical Fragments*, 123.
14   Ibid., 125.
15   Hillman, *The Chemical Muse*, 81.
16   Pliny the Elder, *Remedies from Garden Plants*, Ch. 76.

Specifically, the drug was used for euthanasia on the Greek island of Keos, where those over age sixty were supposed to drink hemlock or poppy when they became "weak or disabled." With the island under siege by Athens around 375 BC and food running out, opium murder or suicide were preferable to slow starvation or slaughter.[17]

\* \* \*

Several centuries later across the Adriatic Sea, a successful case of opium assassination changed the course of Roman history and led to the rise of one of the most vilified leaders of all time.

In 55 AD, a fourteen-year-old prince named Tiberius Claudius Caesar Britannicus was on a fast track to succeed his father, the emperor Claudius.[18] Unfortunately, his stepmother Agrippina had other ideas. She wanted *her* son, the infamous Nero, to be emperor.

There are several conspiracy theories regarding how Agrippina managed to get Britannicus out of the way. According to Tacitus, it happened at a typical Roman feast. Young Britannicus and Nero were, in keeping with the tradition of the royal court, at the kids' table.[19]

---

17   See Kritikos and Papadaki, "The History of the Poppy and Opium in Antiquity"; Parkin, *Old Age in the Roman World;* and Ring, Watson, and Schellinger, *Southern Europe.* Cruel as it may sound, geronticide by drugs is certainly better than being an old Inuit and having "the kids" push you off alone onto an ice floe; or, growing old in a "Dark Ages" Germanic tribe called the Heruli. They are accused of laying their elders on a stack of wood, stabbing them to death, and then lighting the wood on fire. Of course, this was according to the famously self-righteous Roman historian Procopius, who was also almost struck dumb with horror by the fact that their male warriors "practiced" homosexuality, and who accused the wives of Emperor Justinian and his loyal General Belisarius of doing things with geese that we might not want to go into.

18   Claudius isn't as well known as his predecessor and nephew, Caligula, or his stepson and successor, Nero. His reign was fairly benign by comparison—although, admittedly, the bar was set rather low.

19   Hays, "Ancient Roman Food, Spices, and Banquets."

Since poison was frequently used as a weapon at the time, Roman royals took the cautionary step of having their servants taste everything first. At some point during the evening, one of them brought Britannicus a hot drink and took that ritual first sip before serving it. Since he was still standing a few minutes later, Britannicus gave it a try. But first he asked to have some cold water added. That's when someone slipped him an opium Mickey—with a little hemlock mixed in just to make sure it did the job. When he collapsed to the floor, everyone looked suspiciously at Nero, who shrugged, thinking his brother was having just another of his epileptic fits. Agrippina, however, looked on in horror. Tacitus thought that proved she *wasn't* implicated, but later commentators suggested that her horror in the face of behavior that is typical of epilepsy was itself evidence that she was involved and may have been exaggerating her shock—perhaps she was the only person who knew it *wasn't* epilepsy.[20]

As Britannicus left the historical stage, the dinner continued, and within hours he was dead and Nero was directly in line for the throne.

\* \* \*

While most of the early work on opium was done in Ancient Greece, as Britannicus's demise demonstrates, the Romans caught on quickly—not surprising since many Greek physicians found themselves in the Roman Empire after the Battle of Corinth in 146 BC.

One of those Greek-born Romans was a physician and pharmacologist named Pedanius Dioscorides. He spent twenty years—between

---

20   Paratico, "Nero Did Not Murder Britannicus."

50 and 70 AD—writing a five-volume book called *De Materia Medica*. Considered the forerunner of all modern pharmacopeia, it describes the use and abuse of 342 different medicinal plants.

Like Hippocrates, Dioscorides described a few different varieties of poppies. Although he rightly said that only two—the red (field) poppy and opium poppy—were soporifics, others had medicinal purposes, alone or in combination. He knew how poppies were cultivated; he knew how they were harvested; and he knew all kinds of uses for the "liquid that flows out" as well as the best way to prepare and administer it:

- Skin infections: Pound the heads with polenta and use as a poultice.
- Coughs: Boil the heads until the water is reduced in half; add honey; boil again with the honey; add a touch of soothing acacia and hypocistis (an astringent) to help with congestion and drip it on your throat.
- "Crusts of ulcers": Mix with leaves and flowers of the sea poppy and apply directly.
- "Thick or cobweb-like stuff in urine": Boil the root of the sea poppy until it's half the volume and drink.
- Eye problems: Mix it with roasted egg yolk and saffron for inflammation of the eyes.
- Insomnia: "Put up with the finger as a suppository."[21]

Just as physicians today disagree about the risk-benefit equation of opium, Dioscorides thought that physicians of former times were clueless, or at least unnecessarily paranoid, about opium's risks, especially when it came to earaches and eye troubles. He singled out two medicine men from the third century BC for special criticism: Andreas

---

21   Pedanius Dioscorides (of Anazarbos), *De Materia Medica*.

of Karystos, who said that if it wasn't diluted, the sap could make you go blind; and the better-known physician Eristratus, who said it would make your eyes heavy and put you into potentially mortal sleep. He also took a swipe at an ancient named Mnesidemus, who said the only safe way to use it was to inhale but that "otherwise it is hurtful." "These things are false," Dioscorides insisted, "disproved by experience, because the efficacy of the medicine bears witness to the work of it."[22]

Andreas et al. were just a few of the many physicians and pharmacists whose reputations have been tarnished by one surviving reference from a single naysayer. Undoubtedly, they, too, had tried to make sense of their observations of plants and humans' responses to them.

Dioscorides *did* agree with his predecessors about opium's use as a soporific and, thankfully, offered several other means of application besides suppositories. One involved boiling the leaves with the pods, foreshadowing today's use of "poppy straw" to make pharmaceutical morphine.[23] Another prescription involved transdermal delivery— pouring the sap on the forehead and temples.

Dioscorides warned people of the dangers of opium being adulterated—another modern problem that is a 2,000-year-old tradition. He advised that if it were cut with *glaucium* (a non-opium poppy), it would be a saffron color. If "gum" (plant resin, usually from the mastic tree) is added, it would be transparent and lose strength. He reserved his full disgust for people who adulterate the drug by cutting it with grease and setting it on fire to achieve the right consistency

---

22  Ibid.
23  "Poppy straw" is basically the entire plant after the sap has been extracted. Between the alkaloids still in the pod and some others lurking around in other parts of the plant, there's plenty of morphine, thebaine, and codeine left— although there are different extraction methods for each. Poppy straw is the raw material for oxycodone and other narcotics.

and then putting it in a new jar. You can recognize that travesty, he explained, because the end product is softer and more yellowish red.

Little is known about the life of Dioscorides's famous medical contemporary Aulus Cornelius Celsus. His *De Medicina* is a fascinating and encyclopedic book that includes some very down-to-earth diet and exercise advice, as well as precise descriptions of symptoms, their implications for a patient's prognosis, and some immensely practical advice on how to deal with everyday health concerns. For someone with chronic headaches, he advised, "He should not write, read or argue out loud, especially after dinner." In order not to get sick during an epidemic, he said that the best thing to do is "go take a holiday in some distant area," which indicates he had a knowledge or suspicion about the relation between germs and illness that wouldn't get scientifically established until Louis Pasteur appeared in the 1800s.[24]

The poppy appears widely in Celsus's pharmacopoeia, including many historic remedies for insomnia, severe headaches, joint pains, and anal fissures. Rather than alcohol- or water-based concoctions, he frequently recommends making a pill from the ingredients by heating the poppy sap, adding some wine, heating it again, and then cutting it into bean-size pills as it cools. He also gives instructions for making a wide variety of salves that feature "poppy tears."[25]

As useful as he found the poppy in healing, one sentence makes it clear he was also aware of its risks. "The sweeter the dreams," he writes, "the rougher the awakening."[26]

\* \* \*

---

24   Quoted in Richardson, *Celsus on Medicine*, 74.
25   Celsus, *De Medicina*, Book VI.
26   Quoted in Dormandy, *Opium*, 22.

Opium continued its role at the center of Roman royal life during the reign of Marcus Aurelius, a wise, opium-dependent emperor. Just as Alexander the Great ushered in the rise of the Greek empire in spite of (or thanks to) an alcohol-opium psychosis, so Marcus Aurelius bore witness to the fall of the Roman empire while taking regular "maintenance doses" of opium under his doctor's supervision (similar to the maintenance doses of heroin that incurable addicts in some European countries are given today so they can live with their disease and go about their everyday lives.)

Intellectually stoic, emotionally melancholic, physically dependent, and sexually undecided, Marcus Aurelius was a reluctant emperor, an obedient warrior, and a transcendent writer who ruled the Roman Empire from 162–180 AD and is more famous for his philosophical classic *Meditations* than his actions on the barbarian frontier. Aurelius lived in his own world and followed the dictates of his own mind and spirit in the midst of war, revolts, and various personal upheavals, including an unfaithful wife, an unworthy son, and a frivolous co-ruler. Reading his writings today, he seems to have superhuman wisdom, patience, and tolerance. Read in another way, he was a classic addict, living in an illusory narcotic-fueled world to avoid the harsh realities of his life.[27] The importance of these stereotypes falls away, however, in the eternal truths he wrote while on the frontier:

Consider that the things of the present also existed in times past; and consider that they will be the same again. And place before your eyes, from your own experience or from the pages of history, these dramas and scenes: the courts of Hadrian,

---

27    Africa, "The Opium Addiction of Marcus Aurelius."

Antoninus, Philip, Alexander, Croesus; all the same plays, only with different actors.[28]

How did this sensitive soul manage to endure the slings and arrows of his outrageous fortune? How did he manage to maintain his focus, run an empire, and strategize his battles with both the barbarians...and opium...during all those years, without overdosing or getting incapacitatingly addicted? The answer is that his dosage was carefully modulated by one of history's earliest enablers and one of the most important and egomaniacal doctors of all time. His name was Galen.

While Galen admitted that Hippocrates had paved the medicinal way, he believed that he and he alone knew and revealed the true path.[29] "Whoever seeks fame need only become familiar with all that I have achieved," he wrote. While it's true that, to this day, most people know the name Hippocrates better, it is Galen who is referenced again and again by medical historians.[30]

Having learned about opium's painkilling and courage-enhancing properties as a gladiators' physician in his hometown of Pergamon, Galen was able to keep his boss Marcus riding the razor's edge between dependence and addiction, thereby foreshadowing the dilemma that many physicians face today as they are castigated alternately for addicting their patients by prescribing excessive painkillers and failing to alleviate patients' pain by prescribing too little.

Aurelius, like so many others throughout history, may have started taking opium for pain, but soon he was taking regular doses of

---

28   Aurelius, *Meditations*, 110 (Book X, no. 27).
29   Dormandy, *Opium*, 22ff.
30   Quoted in Hajar, "The Air of History."

undetermined strength, dissolved in wine or water with plenty of honey to mask the taste and balance the soporific with the hyperglycemic. This was all well and good when he was back home, but out on the frontier, stressed by weather and warfare, he took increasingly larger doses until he was eating little food while enduring more hardship, raging more eloquently against the cosmic machine, and exhibiting the signs of withdrawal anytime he found himself too tired to do what his stoic nature insisted he do.

Meanwhile, Galen, in addition to modulating Aurelius's daily fix, was formulating countless medicines, many of which bordered so closely on the poisonous that he insisted they shouldn't be given to young children. If high doses were given accidentally, he suggested vomiting and drinking white wine might be an effective antidote.[31]

When not cooking up some new remedy or doing dissections in public (including exposing the laryngeal nerve of a squealing—and then suddenly not-squealing—pig), Galen was writing medical texts—in Greek, just like his famous patient. He produced thousands of pages, many so technical that classicists have thrown up their hands at the idea of translating them with any veracity. Still, more of his words have managed to survive than Plato's, Aristotle's, or any other classical writer, and the ones that medical professionals *were* able to decipher became standard operating procedure until the 1800s, when Galen got a taste of his own presumptive medical ego and physicians and pharmacists even more arrogant than he began to mock his remedies even as they took cues from them.[32]

---

31  The vomiting perhaps might have helped. But alcohol and opioids actually potentiate each other (enhance their effect).
32  The exception was Paracelsus (whom we'll meet shortly), who, in the 1500s, dismissed much of what Galen said.

# Chapter 5

# A Little Light on the "Dark Ages"

Conventional wisdom contends that after the fall of the Roman Empire, opium—and virtually everything and everyone else in Europe—disappeared from the historical record.

Indeed, in the centuries that followed, hordes of barbarians crashed down in waves from the North and East—Goths, Visigoths, Ostrogoths, Huns, and Vandals—moving west and south, expanding like the roots of a tree as they successfully drove out Roman forces and conquered the provinces.

However, even though opioids may go in and out of favor, as long as there's a narcotic juice drying on the outside of some poppy pod, someone is always going to be giving it a taste. While poppy cultivation and use only flourished in the Eastern Mediterranean and Egypt during the "Dark Ages,"[1] some Europeans continued to use it medicinally, in particular Benedictine monks who began reviving the works of Hippocrates and other classic physicians in the late 400s. These monks built monasteries in western Europe and every monastery had an herb garden that would naturally have included the opium poppy.

---

1  A papyrus from the third century AD recommends using opium as a painkiller and includes it as a key ingredient in medicines such as a tea for some unknown condition that was made of opium, beaver musk, and other unspecified ingredients, all diluted with raisin wine. See Saunders, *The Poppy*, 38.

The most famous Benedictine to write about opium was a brilliant abbess (as well as a composer of transcendent melodies and visionary paintings) who lived in the twelfth century. Her name was Hildegard and she grew poppies at her monastery in the small German town of Bingen. She cautioned, however, that opium's dangers were greater than its potential to cure, so it should only be used in the most serious cases.[2]

Modern medicine began to take shape during the Middle Ages, with the appearance of hospitals in Europe, complete with physicians, nurses, and orderlies. Medical schools were also being established in cities such as Bologna, Padua, Paris, Oxford, Cambridge, Prague, and Vienna, where students were able to study opium's usefulness as a painkiller. It was mentioned in several medical texts of the time, including *Premnon Physicon* by the Benedictine monk Alphanus, which is the earliest surviving book in European literature that references it. In the 1200s, the Bishop of Cervia (Bologna) used an anesthetic called *spongia somnifera* that included the usual soporific ingredients: the juice of the unripe mulberry, hyoscyamus (henbane), hemlock, mandrake, wood ivy, mulberry, and poppy juice. The physician mixed them all together, boiled them until the sun set, and whenever needed, the sponge could be put over the nose of the patient during surgery. (If necessary, he could be woken up with a sponge dipped in vinegar or the juice of fenugreek.)[3] This technique, by the way, is very similar to how ether was first administered in the 1800s.

By the end of the Middle Ages, remedies for an anesthetic called *dwale* also began to appear in European medical manuscripts. Formulated to "make a man sleep whilst men cut him," one version instructs the surgeon (likely to also be a barber in those days) to take:

---

2   Schadewalt, "Hildegard von Bingen and the Medicine of Her Time."
3   Dormandy, *Opium*, 40.

three spoonfuls of the gall [bile] of a barrow swine [boar] for a man, and for a woman of a gilt [sow], three spoonfuls of hemlock juice, three spoonfuls of wild neep [bryony], three spoonfuls of lettuce, three spoonfuls of pape [opium], three spoonfuls of henbane, and three spoonfuls of eysyl [vinegar], and mix them all together and boil them a little and put them in a glass vessel well stopped and put thereof three spoonfuls into a potel of good wine and mix it well together.[4]

Patients would feel no pain—occasionally because they died of an overdose.

In addition, there continued to be folk healers in rural areas—women and men whose huge pharmacopeia included virtually every animal, vegetable, and mineral substance in the vicinity, depending on the season. These healers often found themselves in a precarious position. Their livelihoods depended on people believing they possessed secret knowledge. However, if their remedies worked, they could be accused of being witches; while if they didn't work, they could be damned as charlatans.

Regardless, as the knowledge and use of opium continued quietly in Europe's Dark Ages, a few hundred miles to the southeast one of the most medically advanced ages in history was about to bring the palliative and perilous promise of opium powerfully back into the light.

---

4   Carter, "Dwale."

# Chapter 6

# Opium's Golden Age

By the 800s AD, a series of Persian physicians began to evolve the principles of Greco-Roman medicine into theories and practices that resound to this day. They were all philosophers and polymaths, and all made major contributions to the ever-growing body of knowledge about opium's medicinal properties. In addition to using it to treat pain, they explored ways to standardize drug potency and mitigate the effects of drug tolerance. They also provided descriptions of drug withdrawal that could have been written today.

Thabit ibn Qurra (836–901) was a Kurdish hermetist who developed a cough syrup of rosewater and grape syrup with opium, a thousand years before the A.H. Robins Company figured out how to isolate codeine from opium poppies to make Robitussin.[1]

Even more famous was Abu Bakr Muhammad ibn Zakariya al-Razi (or Rhazes, ca. 865–925), who pioneered neurosurgery and ophthalmology. Considered the father of pediatrics, he wrote 200 books including a twenty-three-volume encyclopedia of anatomy and diseases, as well as the first medical manual for ordinary folks to

---

1   "A.H. Robins Company."

use at home. In addition to the usual opium remedies for pain, he recommended applying it topically, with henbane or hemlock, for inflammation of the joints and gout.[2]

Abu al-Qasim al-Zahrawi (936–1013) wrote the first illustrated book about surgery, invented several surgical instruments, and wrote about using opium as an anesthetic.[3]

The great doctor and philosopher Moses Maimonides (1135–1204) spent much of his life in Egypt as the personal physician to the famous Islamic sultan Saladin. In his off-hours, he treated the sick people who crowded in the courtyard outside his home—rich and poor, Muslims, Christians, and Jews. Long before Cicely Saunders came up with the idea of hospice, Maimonides was giving opium to the dying whether they could afford it or not because "the dying must not be left to suffer."[4]

Then there was a woman (or several women working together) in twelfth-century Salerno, Italy, who, writing under the name "Trota," published a three-volume collection called *The Trotula*. It dealt primarily, but not exclusively, with obstetric and gynecological medicine, providing advice on menstruation, pregnancy, caesarian section, and childbirth. Trota strongly believed in using opium to alleviate the pain of childbirth and advocated a lifestyle that is remarkably similar to what doctors all over the world prescribe today—balanced diet, exercise, cleanliness, and herbal remedies.[5]

Still, in terms of opium knowledge—personal and professional— we can learn the most from the life of the Persian physician known as Avicenna (980–1037).[6] If Rhazes was one of the most brilliant medical

---

2    Tibi, "Al-Razi and Islamic Medicine in the 9th Century."
3    Flascha, "On Opium."
4    Dormandy, *Opium*, 35.
5    Lewis, "A Surprisingly Long List of Medieval Women Writers."
6    One sign of his status is that he made it into the first ring of Dante's Hell where he joined other medical luminaries including Dioscorides, Hippocrates, and

minds in history, Maimonides the paragon of the selfless physician, and Trota the first female (and arguably feminist) gynecologist, then Avicenna surely ranks as one of the most brilliant, tormented, and irreverent free spirits the medical world has ever seen.

Avicenna was born in Afshana near Bukhara in modern Uzbekistan, the "Shining Pearl" of the so-called "Silk Road" (which has also always been a spice and opium road). According to no less an authority than himself, he amazed his father and the other local wise men by memorizing the Qur'an by 10, studying medicine by 13, and taking on patients by 16. The next two stops on his autodidactic educational fast track was an internship at a hospital in Baghdad and the successful treatment of a big-time Baghdad ruler, which gave him access to the royal library. A Renaissance man who lived 500 years before the Renaissance, Avicenna wrote more than 400 works including two astounding encyclopedias. One covered the sciences, including psychology, geometry, astronomy, math, and music. The other, known as *The Great Canon*, has been called the "most famous single book in the history of medicine."[7]

Avicenna wasn't the first person to spend his life trying to reconcile his religious beliefs with what he saw with his own two eyes. But he did it with a fierce, unrelenting, and uncompromising focus. His goal was nothing less than a grand vision that reconciled the existence of good and evil, prophecy and miracle, mosque and state, philosophy and physics. He had no patience with people whose understanding he considered superficial or religious leaders who accepted things

---

Galen. Dante admired them all, but reluctantly accepted the fact that, since they'd never been baptized, they would be in Limbo forever. Still all three physicians were in excellent company, joined by Plato, Socrates, Homer, Julius Caesar, and Brutus.

7   O'Connor and Robertson, "Avicenna Biography."

that appeared magical without question or examination, and no one was safe from his critical pen. (He even dismissed Rhazes's work, suggesting he should have stuck to studying urine.)

While engaged in this intense inner questioning, Avicenna's outer life was one long walk on a wild side. It included political intrigue, imprisonment, going into hiding, and riding into battle with his patron prince as the Samanid dynasty fought for its survival.

Between personal experience and his careful observation of the patients for whom he prescribed it, Avicenna was one of the foremost opium experts in history. He recognized the need for a uniform standard for opium (much as the medical community and dispensaries are struggling to establish standard replicable doses of cannabis today). The one he developed, based on the size of chickpeas, is still used in some Muslim communities from Morocco to Indonesia.[8] He also created standards for how poppies should be harvested and processed in an attempt to mitigate variations in potency and, in a warning that resonates today as street heroin is laced with even more deadly fentanyl, he cautioned physicians to gauge a patient's tolerance carefully and urged people to buy only from reliable sources.

Even if the sources *were* reliable and the price was right, he acknowledged that opium (*Afion*, they called it then) came with its risks, from constipation to overdose. In cases of the latter, and maybe the former, he recommended that you keep the patient walking, talking, and vomiting.

His medical recommendations involved using opium in various ways—edibles, extracts, tinctures, suppositories, and ointments, as well as just dissolving it in wine and sipping only as directed. In some

8    Latimer and Goldberg, *Flowers in the Blood*, 37.

cases, he even recommended scarifying the skin and applying it—a herald of the eventual development of hypodermic administration.[9]

In addition to the basic applications of opium that physicians had written about since ancient times, Avicenna found that opium had benefits in treating many other conditions. While he did not directly discuss modern concepts such as the "central nervous system" or "opioid receptors," he recognized that opium worked by preventing "nerves to conduct painful sensory impulses."[10]

Given the choice, he believed in treating the *cause* of pain and not using opium unless absolutely necessary. He was also aware of the consequences of opium abuse: memory and reasoning dysfunction, difficult breathing even when applied topically to the chest, muscle spasms, dyspepsia, and sexual dysfunction.

Another indication of the precision of his observations is that he identified more than a dozen signs of opium overdose and how alcohol could magnify the danger. Regardless, once a person developed respiratory distress, cold breath, and cold sweating, he said death was almost sure to follow.

Avicenna made two other very different observations that are particularly trenchant, even poignant—one about others, and the other about himself. The first is that, in general, patients seek out medical care for two reasons—pain and fear. He said fear was worse, but opium could relieve both. Second (like many a risk taker), he said he wanted to live his life "in breadth, not in length."[11]

Fittingly, Avicenna spent his final days in his hometown of Afshana,

---

9    He even claimed to have seen a patient die from too high a dosage of an opium suppository.

10   Heydari, Hashempur, and Zargaran, "Medicinal Aspects of Opium as Described in Avicenna's Canon of Medicine."

11   Dormandy, *Opium*, Ch. 5, n7.

which was notable for its lush fields of poppies—which worked out well for him since, on top of his many other obsessions, he was a serious user. While some suggest that Avicenna died from a botched self-treatment after becoming ill during a battle, others argue that he may have been poisoned by a servant. In either case, the physiological cause was an overdose of opium.

\* \* \*

As the Middle Ages wound down, the first indications of a true societal drug crisis appear in the writings of a physician named Hakim Imad al-Din Mahmud ibn-Mas'ud Shirazi (1515–1592), who participated in, chronicled, and analyzed an alarming increase in the recreational use of opium in Persia. Among his extensive medical writings, he describes how to use ginseng to treat opium addiction.[12]

Hakim Imad al-Din learned medicine from his father. His first job was as court doctor for the ruler of Shirvan, a region in modern Azerbaijan. Early in his career, he did something that infuriated his patron. We don't know what that "something" was. He was a young man, so perhaps it was a misdiagnosis, a botched treatment, or even an ill-advised romantic entanglement. Whatever it was, his boss decided on a punishment somewhere between a slap on the wrist and being stoned to death: Hakim Imad al-Din was forced to spend the night outside in a freezing pool. In order to withstand the pain, he passed the long night eating opium. While he survived, he never stopped shivering—he lived the rest of his life with a severe palsy. He also

---

12    Golshani, "Hakim Imad Al-Din Mahmud Ibn-Mas'ud Shirazi (1515–1592), a
      Physician and Social Pathologist of Safavid Era." NB: Recent studies show that
      ginseng also may have some effect on STDs. See Penn State Medical Center,
      "Sexually Transmitted Diseases."

became an opium addict, which led him to write the first standalone work on opium called *Resaleh Ophioun*, or *The Book of Opium*, in which he talked about the chemistry, uses, and abuses of the drug.

An impressive number of Hakim Imad al-Din's contemporaries in sixteenth- and seventeenth-century Iran were heavy, frequent users. Their staple consisted of small pea-sized balls of opium that they swallowed like pills. But they also smoked it along with hashish and, when they partied, drank poppy seed tea along with a variety of alcoholic beverages. (Opium use was and is permitted in Islam and the prohibition on both alcohol and hashish was more honored in the breach—especially among the young.) In fact, rulers were said to give opium to their children for their collective amusement and even to curtail their ambition in order to prevent them from plotting to overthrow their fathers.

Hakim Imad al-Din was likely an important source for the descriptions of tolerance, addiction, and withdrawal that were written by a contemporary French diamond merchant named Tavernier, who traveled extensively through the East. His words reveal the toll that this early opioid crisis was taking: "Opium addicts admitted that opium was harmful, but they said that it was their habit. When they did not take opium, they had a pale-yellow face and were always weak and sleepy. They needed greater and greater doses of the substance to achieve the same original effect. The effects of a dose of opium last for about four hours. Then, the body rebounds with a set of withdrawal symptoms; the symptoms include watery eyes, muscle pain, anxiety, agitation, nausea, and insomnia." In fact, Tavernier reported, the first thing they did when they broke their Ramadan fast was prepare their hookahs.[13]

Opium also contributed to the other cultural scourge that was

---

13   Tavernier as quoted in Golshani, "Hakim Imad Al-Din."

spreading across Europe and the Middle East at the time—syphilis. The drug is, in some forms and dosages, used as an aphrodisiac and back then it was often taken mixed with wine, which, like any alcohol, lowers inhibition: the perfect formula for spreading STDs then and now.

There was a lot of finger pointing during the late Middle Ages about who was responsible for the spread of syphilis in Europe—a pattern of blame that would prove to be part and parcel of every opioid crisis to come. The argument began with the question of whether Columbus and his crew brought the disease back from the Americas or if it was in Europe all along, just waiting for a diagnosis. Presuming the latter, the Italians called it the French disease, the French called it the Italian disease, the Dutch called it the Spanish disease, the Russians called it the Polish disease, and the Turks called it the Christian disease. It doesn't seem anyone called it the Persian disease, but they suffered from it as much as anyone.

# Chapter 7

## The Monarch of Medicine

While opium remained a major ingredient in Persian medicine throughout the Dark Ages, it wasn't popular in Europe until the eighteenth century, when an opium-based tonic called laudanum led to a wave of addiction in Europe and America. Laudanum had first been introduced to the Western world in the sixteenth century by a cantankerous and absurdly talented physician, philosopher, astronomer, astrologer, theologian, alchemist, and opium-eater named Paracelsus (1493–1541).

With the possible exception of Mary Shelley's fictional Dr. Frankenstein, there has never been a physician so willing to explore the outer reaches of the chemical universe in search of the secrets of health and eternal life. (In fact, Frankenstein found great inspiration in Paracelsus, whose "wild fancies" he studied "with delight").[1]

Paracelsus was the epitome of sui generis. Some admirers even referred to him as the Martin Luther of medicine. For his part, ever the heretic, he once described Luther and the pope as "two whores discussing chastity."

Paracelsus's insistence on liberating medicine from the shackles

---

1    Monahan, *They Called Me Mad*, Ch. 2.

of Greco-Roman dogma was perfectly in accord with other great minds of the Renaissance: Columbus helped prove that the earth wasn't flat; Copernicus argued that the same earth wasn't the center of the universe; Vesalius exposed the flaws in Galen's theories of anatomy, transforming how we see ourselves; while Leonardo da Vinci revolutionized how we see virtually everything else. Even if his name isn't as familiar as those of these extraordinary contemporaries, Paracelsus was equally influential. Aided and abetted by Gutenberg's invention of the printing press (which facilitated mass production of their works), all their insights and inventions transformed science, medicine, religion, and art.

Paracelsus spent much of his life traveling—by choice and necessity. His modus operandi was to arrive somewhere, perform a miraculous cure or two, and, with alarming regularity, annoy the local physicians, pharmacists, and town officials with his personality and medical theories to the extent that he found himself compelled to hit the road again. He wandered in this way throughout Europe until, in 1521, at age twenty-eight while living in Moscow, he was swept up in an invasion by Tartar warriors who took him to Crimea as a prisoner. Before long, however, his captors recognized his medical skills and brought him along on a diplomatic mission to Byzantium.

Paracelsus's conviction that he knew a lot of things stuffy European physicians were clueless about was due in large part to the time he spent in the Middle East. There, he was exposed not only to the encyclopedic knowledge of the Islamic physicians but also, as he put it, from "dervishes in Constantinople, witches, gypsies, and sorcerers, who invoked spirits and captured the rays of the celestial bodies in dew." From them, he claimed to have learned how to cure the incurable, give sight to the blind, and wake the dead by giving them an opium brew so powerful and foul tasting it may have woken the dead all by itself—the one that he called laudanum.[2]

Although users had been dissolving opium and various other ingredients in alcohol for years, and would for years to come, Paracelsus is widely considered the person who popularized laudanum. One reason is that he made it more palatable than most alcohol-opium solutions. His esoteric formula was:

Take of Thebaic opium, one ounce; of orange and lemon juice, six ounces; of cinnamon and caryophilli, each half an ounce. Pound these ingredients carefully together, mix them well, and place them in a glass vessel with its blind covering. Let them be digested in the sun or in dung for a month, and then afterwards pressed out and placed again in the vessel with the following: Half a scruple of musk and half a scruple each of the juice of corals and of the magistery of pearls. Mix these, and after digesting all for a month, add a scruple and a half of the quintessence of gold.[3]

While some of these instructions and ingredients may have had little to do with its effectiveness, both pearls and gold do have medical applications.[4] More significantly, the acid in the orange and lemon juices would have changed the molecular structure of the morphine in the opium so the body could metabolize it more efficiently. This is a principle that was central to the synthesis of heroin from morphine

2   Jaffe, *Crucibles*, Ch. 2.
3   Paracelsus and Laurence, *Hermetic Medicine and Hermetic Philosophy*, 62.
4   See, for example, Bahn et al., "Control of Nacre Biomineralization by Pif80 in Pearl Oyster"; and King, "Uses of Gold in Industry, Medicine, Computers, Electronics, Jewelry."

centuries later.[5] Never one to be modest, Paracelsus insisted that his formula was far more powerful than opium alone and could free patients from the pain caused by any disease—which, in the sixteenth century, included many painful diseases indeed. From his perspective, while not necessarily a cure-all, opium's calming and soporific qualities were themselves a critical part of healing.

Other physicians of the time shied away from using laudanum either because they were less impressed by its power, more fearful of its dangers, or in some cases determined to reject anything created by this medical apostate. There was another crucial reason—this one political—for many Europeans' disdain for laudanum: with the Inquisition in the air, some doctors may have been worried about being associated with opium because it came from the Islamic world.

Paracelsus, however, dismissed all their fears and opinions. During his brief stint as a teacher at the University of Basel, he told his fellow professors that: "You are nothing but teachers and masters combing lice and scratching. You are not worthy that a dog shall lift his hind leg against you. Your Prince Galen is in Hell, and if you knew what he wrote me from there you would make the sign of the cross and prepare yourselves to join him. Your dissolute Avicenna, once Prime Minister, is now at the gates of Purgatory."[6] Ten years after Luther nailed his contrarian theses to the door of All Saint's Church, Paracelsus doubled down during the celebration of Saint John the Baptist, which always took place on the summer solstice. The festivities, which had roots going back to ancient pagan celebrations, involved lighting bonfires to ward off evil spirits. While there are disputing accounts of what

---

5    Latimer and Goldberg, *Flowers in the Blood*, Ch. 3.
6    Quoted in Jaffe, *Crucibles*, 18.

happened in Basel, Switzerland, on St. John's Day in 1527, the event is a seminal moment in medical history.

Some say Paracelsus staggered, drunk as well as stoned on opium, onto the Basel village green, where he added the works of Celsus and Avicenna to the traditional bonfires. Others say he was determinedly sober and walked into the crowd holding a large brass vase stuffed with ancient books and papers—Pagan, Muslim, and Christian medical doctrines—which he proceeded to ignite wholesale.[7]

In any case, as the crowd of horrified professors and cheering students looked on, he cried: "Ye physicians, so-called, of Paris, ye of Montpellier, ye of Swabia, ye of Cologne, ye of Vienna, ye of Meissen [and many others] and those who dwell on the Danube and the Rhine, ye islands of the sea, thou of Italy, thou of Dalmatia, thou of Sarmatia, thou of Athens...ye Greeks, ye Arabs, ye Israelites, after me and not after you! Even in the remotest corner there will be none of you on whom the dogs will not piss. But I shall be monarch and lead!...So gird your loins and forget the past!"[8]

He was promptly driven out of town.

Toward the end of his short life, presumably unaware that his medical legacy would outlive that of the professors he disparaged, Paracelsus spent time in several German cities, where, among other things, he tried treating syphilis with an ointment that featured minuscule doses of mercury. For, as he put it, "All things are poison and nothing (is) without poison; only the dose makes that a thing is no poison."[9]

---

7   Another writer claims: "His first official act at the University [of Basel] was to build a bonfire in his lecture room to burn the works of Galen." The same source says that his enemies spread the rumor that he hadn't married because in his youth he'd been castrated by a hog. Haggard, *Devils, Drugs, and Doctors*, 347.
8   Quoted in Dormandy, *Opium*, 46.
9   "Paracelsus."

Five centuries later, when the difference between palliative and poisonous dosages of synthetic opioids can be as small as a speck of dust, these words are more relevant than ever.

Paracelsus was an alchemist—the most famous of all time, but he was by no means a quack or even a self-interested entrepreneur. While he would have been more than happy to find a "Philosopher's Stone"—the mythic chemical most often associated with alchemy, that could supposedly transform base metals into gold—the gold that he truly sought was human health and vitality. The "base metals" that he believed were needed to attain it included now familiar supplements like iron and zinc as well as powerful chemicals like mercury and sulfur.

Paracelsus lived at a time when the rediscovery of both the classics and ancient mystical texts (*Corpus Hermeticum*) were pulling medicine in different directions: one that we call allopathic medicine, which has traditionally focused on curing symptoms; and the other, which we now call alternative or mind-body medicine, which focuses on treating the whole person. His ability to synthesize the two many centuries ago has, for the most part, been forgotten. But whether they realize it or not, every time a physician recommends that a patient meditate as well as medicate; or a researcher develops a clinical trial for a drug that could prove to be poison or cure; or an herbalist acknowledges the need to include the judicious use of narcotics to deal with severe pain; or, most significantly, a terminal patient is given morphine to reduce suffering, he or she has returned to the path of Paracelsus.

# PART III

## Opium Goes Global

# Chapter 8

# Marco Polo
# and the Rise of Global Commerce

Before the modern era, the story of opium described the independent rise and fall of different civilizations—Neolithic, Sumerian, Greco-Roman, Persian, et cetera. In the late Middle Ages, however, it began to go global. Throughout the Americas, Africa, Europe, and Asia, buyers, dealers, doctors, missionaries, enforcers, sailors, users, and countless other bit players began competing with each other for dominance in the worlds of finance, medicine, power, philosophy, faith, and fame. Small trading centers grew into major population centers. Others fell into obscurity, like modern cities bypassed by today's interstate highways. Borders expanded, contracted, and even imploded as countries and dynasties used their trading wealth to conquer new lands and forge new alliances, some of which crumbled in the face of changing priorities, threats, and opportunities.

For the most part, the names of the intrepid explorers and traders who first united the cultures of East and West are forgotten, buried in the archives of elegantly detailed manuscripts and commercial contracts—with one notable exception: Marco Polo, a thirteenth-century Italian explorer who wrote himself into the history books as one of the most famous travelers of all time.

Many years after this adventurer returned home, a German geographer named Baron Ferdinand von Richtofen conflated the many routes Polo and other travelers had forged to and from Asia under the singularly misleading name "the Silk Road." There was in fact neither just one road, nor were those who traveled them interested only in silk.

Even more surprising, the routes weren't necessarily started by merchants from the *West* trying to access valuable commodities such as silk from the *East*. Rather, if one person could be credited with opening the "Silk Road," it would be the visionary and expansionist Chinese emperor Han Wu-ti, who ruled more than a thousand years before (r. 141–87 BC) and sent explorers on missions from China to the *West* in search of new herbs, scientific knowledge, and philosophical ideas (which the Chinese valued as much as any commodity).[1]

Since Marco Polo was the first European to *write* about the journey, he is usually credited with being the first European to reach China, even though explorers and merchants had been going there on and off for centuries. However, as a Venetian living when the independent city-states of Genoa and Venice were competing for dominance in East-West trade, Polo was in the right place at the right time to make a name for himself.[2]

That economic competition, however, eventually evolved into full-out war—actually four wars during the 1200s and 1300s—none of which gave either side a conclusive victory. During the second war, Polo returned home after twenty-six years in the East and, eager to continue his life of adventure and fulfill his patriotic duty, commanded himself a small galley, equipped with catapults called trebuchets,

---

1   Others credit his Western contemporary Mithridates II with opening the "Silk Road" from the West to East at virtually the same time. See Kienholz, *Opium Traders and Their Worlds*, vol. 1, 55.

2   The key port for sea trade to the East was in Constantinople, relatively equidistant from the two city-states.

which could sling large rocks or flaming missiles at enemy ships. Un-fortunately, he and much of the rest of the Venetian fleet ran aground off the Croatian island of Curzola. This led to a particularly violent battle during which the Genoese captured 84 Venetian galleys, and 8,000 men.[3]

But Polo's military disaster and subsequent arrest would turn into a literary windfall for civilization. He spent a year imprisoned in the Palazzo di San Giorgio,[4] where he passed much of his time entertaining his roommate, a writer named Rustichello da Pisa, with tall tales from the East. In the process, da Pisa became an early practitioner of the "as told to" biographical style—complete with all the brazen exaggerations and negligible fact-checking those books are known for. The result is the most famous travelogue of all time: *The Book of the Marvels of the World*, better known as *The Travels of Marco Polo*,[5] a book devoted to everything Polo saw (or said he saw) and heard (or said he heard) during his years abroad, from 1271 to 1295.

Everyone who sits around and tells his or her friends highly embellished stories of their high-school or college days can consider themselves literary descendants of Polo...especially since, in both cases, their imaginations may have been fueled in part by a narcotic derived from hemp and/or poppy. Regardless of their accuracy, Polo's stories captured the fascination that the East and West had long held for each other. Indeed, there were enough intriguing details to

---

3   There's some disagreement as to which battle/year it was—It was either 1296 or 1298. Regardless, Polo found himself at the wrong place at the wrong time. See Bergreen, *Marco Polo*, Prologue.

4   His "prison" was a stately mansion repurposed to hold captured noblemen. It resembled the type built in America for white-collar embezzlers far more than the overcrowded federal penitentiaries used to house small-time dealers and addicts.

5   There are many versions and translations of the book. Our quotes are from one of the earliest: Marsden, *The Travels of Marco Polo*.

convince Christopher Columbus it was worth taking Polo's book on his voyages in an effort to study up on the continent he would never reach. The notes he made in the margins of his well-worn copy make it clear that he was determined to capitalize on the riches of pepper, cinnamon, and cloves that Polo had written about.

Although separating fact from hyperbole in Polo's stories can be difficult, a number of anecdotes illustrate how he came into contact with opium and the role it likely played in his life—in particular, two tales from early in his first journey, which take place in Afghanistan and Iran, the center of poppy cultivation today.

Polo's journey began when he was a teenager. His father Niccolo and uncle Maffeo had just returned from a fourteen-year trading mission through Russia and Kazakhstan, across the vast Takla Makan and Gobi Deserts all the way to the court of Mongol king Kublai Khan in Khanbaliq (modern Beijing).[6] Arriving back in Venice, Niccolo and Maffeo learned that Marco's mother had died, so they decided to take the wide-eyed boy with them on their next adventure three years later.

Early in the journey, Polo writes about spending time in a town in northern Persia south of the Caspian Sea, where he heard about a ruler who is described in a chapter called, "Of the Old Man of the Mountain. Of his Palace and Gardens."

The story revolves around the legendary Aloadin, ruler of a small Shiite sect known as the Ismailis. Polo tells us that he and his people lived in a beautiful valley with luxurious gardens, elegant palaces,

---

6   Kublai Khan ruled from 1260 to 1294. At its peak, his Mongol empire was the largest continuous land mass under one rule in history, stretching from China's Pacific shores all the way to the Mediterranean. While it briefly interrupted Chinese dynastic rule, it facilitated, among other things, more rapid exchange of ideas with the West.

and streams of wine, milk, honey, and pure water flowing in specific directions. Continuing in his signature rhapsodic vein, he describes "elegant and beautiful damsels, playing upon all sorts of musical instruments, dancing, and especially those of dalliance and amorous allurement."[7]

Aloadin insisted that nobody enter his kingdom without permission. To make sure, he built an army of brave young men and used opium to guarantee their allegiance. As Polo writes, Aloadin

> entertained a number of youths, from the age of twelve to twenty years from the inhabitants of the surrounding mountains, who showed a disposition for martial exercises, and appeared to possess the quality of daring courage. To them he was in the daily practice of discoursing on the subject of the paradise announced by the Prophet, and of his own power of granting admission to that Paradise. And at certain times he caused opium to be administered to ten or a dozen of the youths; and when half dead with sleep he then had conveyed to the several apartments of the palaces in the garden.[8]

The boys woke up to find themselves being fed and caressed by those singing damsels. They continued enjoying the pleasures of this paradise for several days, at which point they were drugged again and brought out to be interrogated by Aloadin about where they thought they'd been. Even though they were convinced they'd tasted the paradise that awaited them after death, he told them they hadn't seen anything yet. But first they had to prove their devotion to him by

---

7   Wright, *The Travels of Marco Polo The Venetian*, 75.
8   Ibid.

demonstrating their willingness to die in his service. After what they'd been through, it must have felt like they had nothing to lose.

Thanks to their loyalty, anyone who invaded Aloadin's kingdom was quickly dispatched. The boys were known as "hashishis" or "assassins," who, like soldiers throughout history, turned to drugs to give them the courage to do what they were trained to do and later find a way to forget what they had done.

The most compelling proof that Polo himself indulged in opium—medicinally if not recreationally—comes from his oddly brief description of an illness he suffered while in Balashan (Badakhshan) in northeast Afghanistan.

Polo begins, as usual, by describing the wonders of the land—for example, the fabulous mines of silver, lapis, rubies, copper, and lead and the breed of noble fast horses said to be the descendants of Alexander's Bucephalus, with hooves so hard they didn't need shoes, and the ability to gallop full speed down rocky hills. He goes on to describe how the air at the top of the mountains is uncommonly pure and healthy and how whenever the locals get sick, they ascend those hills for three or four days to recover, and that he did the same for most of a year.

The details of his illness indicate it was tuberculosis, which was common in Europe at the time, especially for exhausted traders. Back then, the primary treatment for TB was opium. Since Badakhshan was a center of poppy farming at the time, Polo would likely have used it during his year of retreat and recovery. As one biographer put it:

> One explanation for the unusual length of time that he languished in Badakhshan could be that, in the course of trying to recover from a febrile illness, he became dependent on opium and had to detoxify—a protracted and agonizing process. The

symptoms of withdrawal that he might have suffered include nausea, sweating, cramps, vomiting, diarrhea, depression, loss of appetite, anxiety, and rapid changes in mood. He would have become edgier, moodier, more sensitive to light, and more highly suggestible. Where his father and uncle saw a road or a bridge or a storm, Polo might have seen evidence of impersonal cosmic forces at work, sweeping them toward an inchoate destiny.[9]

Others go even further out on a limb and imagine that Polo's defeat during the disastrous Battle of Curzola stemmed in part from an opium haze. In such a state, Polo failed to give his sailors instructions coherently enough to keep them from beaching their vessel; meanwhile they prematurely celebrated victories over the Genoese.[10]

Even if there is little evidence for the common assumption that Polo brought opium to China, it is likely that he was a user who sourced his drug along the very routes where opium and heroin begin their journey around the world today.

---

9   Bergreen, *Marco Polo*, Ch. 4.
10   "This Day in Alternate History: Guest Post: Marco Polo Introduces Opium to the West."

# Chapter 9

# "The Spice Trade Was in Reality the Drug Trade"

What's remarkable about the Age of Exploration is that one of its most powerful drivers was the search for spices, including mind-altering botanicals such as opium and cannabis. Indeed, for centuries, sailors and traders were willing to spend months traveling on barely charted seas, risking life and limb, killing and capturing and being killed and captured by peoples whose languages and cultures they didn't understand, merely to buy and sell some pepper, cinnamon, cardamom, ginger, turmeric, and nutmeg, all of which are so common today.

To be sure, spices had a number of important uses: they disguised the smell of rotting meat, cleared the air from the smell of death during plagues, were burned as incense in religious ceremonies, were a sign of wealth among the nobility, and, in some societies, considered aphrodisiacs.[1]

However, the most important reason that Europeans were willing to pay so dearly for spices was, as Yale professor Howard Haggard noted in the early twentieth century, "The spices that were sought

---

1   For example, in the twelfth century, Duke William of Aquitaine claimed he achieved a total of 188 "exertions" during a weeklong ménage à trois, thanks to consuming a sinful amount of pepper. See Turner, "The Spice That Built Venice."

were used in medicine rather than as condiments: the spice trade was in reality the drug trade."[2]

Medical treatments often involved a combination of spices with natural narcotics and stimulants. For example, in addition to opium for pain, Paracelsus's famous laudanum included both cinnamon and clove. Cinnamon and its derivatives are antioxidant, anti-inflammatory, and antimicrobial. They have also shown promise in the treatment of Alzheimer's, diabetes, arthritis, arteriosclerosis, and cancer. Clove is another powerful antioxidant that has antimicrobial, antiviral, and analgesic properties.[3] Modern pharmacology still takes advantage of herbal remedies from this time: camphor remains popular for treating respiratory ailments, aloe for skin irritations, rhubarb for digestion, sandalwood as an antiseptic and for circulatory health, and, most recently, the use of cannabis extracts for a wide variety of conditions.

In addition to medicinal herbs, other narcotics and stimulants began to play a major role in world trade during the 1500s: tobacco from North America, coca from South America, hemp from the Middle East and India, beer from Germanic countries, rum and hard spirits from the British Isles, wine from southern France and Italy, coffee from Arabia, sugar from Persia and India, and tea from China. All contributed to the growth of worldwide trade. People became dependent on them all. But opium was the one that users couldn't do without.

\* \* \*

---

2   Haggard, *Devils, Drugs, and Doctors*, 332.
3   These descriptions are not from books on natural medicine. They can be found on National Institutes of Health websites: https://www.ncbi.nlm.nih.gov/pmc/articles/PMC4003790/ and https://www.ncbi.nlm.nih.gov/pmc/articles/PMC3819475/.

By the beginning of the 1500s, shipping technology was improving by leaps and bounds; boats were getting larger, sturdier, and faster; navigation had become more reliable; and long ocean voyages were finally possible. In a relatively short time, those routes began to supplant traditional overland routes. Therefore, maritime prowess, and the number of trading posts it enabled a country to control, became a far better indicator of power and wealth than a country's size or population. Every coastline in the Far East was soon dotted with European settlements. At the same time, governments and traders were developing new modes of financing, while building an evolving network of both formal and informal trade rules and regulations.

The Portuguese were the first European sailors to dominate worldwide sea trade. In 1488, while Christopher Columbus was still trying to figure how to get to the East by sailing west, a Portuguese nobleman named Bartolomeu Dias began charting an entirely different ocean course. His three-ship expedition sailed south—down the Atlantic, along the west coast of Africa, around the tip and then east into the Indian Ocean. By then, he was convinced India was within reach, but his crew was exhausted and only wanted to return home. Fearing an impending mutiny, Dias merely erected a stone cross at Kwaaihoek, on the east coast of South Africa, and went back, naming the tip the Cape of Good Hope along the way.

Less than a decade later, in April 1497, a fellow Portuguese adventurer named Vasco da Gama took over where Dias had left off, rounding the Cape of Good Hope with an "armada" of four ships before then sailing across the Indian Ocean to Calicut on the west coast of India. When he arrived there, a full year later, he was disappointed to find that the Moors, Muslim traders who also came from the Iberian Peninsula, were already there, having traveled by the traditional overland routes. (The Portuguese Christians needed a sea route to avoid the danger of traveling through the Islam-controlled Middle East. The Moors, who were Muslims, didn't have that problem.)

The Muslim traders, in turn, were shocked to see da Gama arrive by boat—especially from the largely uncharted southwest. Da Gama sent a Spanish-speaking crew member named Joao Nunes to make his boss's mission perfectly clear: "We seek Christians and spices." Nunes's implied threat was equally clear: his boss had every intention of establishing a Portuguese monopoly on trade in India, converting as many natives as he could, and he would literally take no prisoners in the process.

While da Gama developed passable relations with the local Hindus, he failed to drive out the Muslims and, after three months, short on manpower and supplies, he turned back. (Even though only 54 of his 170-member crew survived the grueling adventure, many back home were surprised any had made it at all.)

The next year Pedro Álvares Cabral led a heavily armed fleet of thirteen ships and 2,500 men on the same route. He lost four ships rounding the Cape of Good Hope but eventually made it to Calicut, where he established a warehouse. When it was destroyed in a riot, he bombarded the city before going 150 miles or so further south and setting up a new warehouse in Cochin.

In 1502, determined to establish Portugal's presence once and for all, da Gama returned with twenty ships, more than a thousand men, and his signature ruthlessness. In the spirit of the Inquisition, he spread fear and loathing among any non-Christians he happened upon. Once called "a fiend in human form," he blew up a merchant ship with 700 Indian Islamic converts who were on their way back from a pilgrimage to Mecca. He also burned down a dozen or so Muslim merchant ships and butchered their crews. This time, he had no qualm of treating native Hindus with equal efficiency—gouging out the eyes and hacking off the noses and hands of the unfortunate prisoners who were sent by the king of Calicut to keep him from setting up shop. (He returned the appendages to the ruler with the suggestion that he make a curry out of them.)[4] By the time da Gama was done, he had made it utterly clear that if you wanted to trade in India, you had to be a Christian and have a Portuguese permit, otherwise he would take both your merchandise and your life.

Having proven that military dominance was the best trading strategy, the murderous sociopath died around age sixty and was buried in India, but, concerned that his body might be torn limb from limb by revengeful natives, the Portuguese king had him exhumed and brought back to Portugal.[5]

In the years to follow, most Portuguese focused on spices rather than the kind of senseless slaughter that had made da Gama famous.[6] While not above waging attacks on the locals to make a claim, they often found it more advantageous to form alliances with native rulers. In the first two decades of the 1500s, a dozen or more Portuguese forts and settlements appeared in rapid succession along the coasts of India, Sumatra, Borneo, and other territories in Asia.

In terms of the opium trade, the most significant Portuguese voyage was that of Afonso de Albuquerque in 1511. He took strategic control of the Strait of Malacca—a shortcut between the Indian Ocean and the Pacific that, being narrow and easy to guard, helped protect Portuguese claims to the East. He reported that he saw opium sold in Burma, the Malay Peninsula, and later in China. In fact, he was so

---

4    Robins, *The Corporation That Changed the World*, 43.

5    Da Gama's empty Indian grave became a popular tourist site, and there's a large mural of him in the lobby of India's Spice Board in Cochin.

6    Not everyone would agree that da Gama was an exception. Economic historian Niels Steengaard writes, "'The principal export of pre-industrial Europe to the rest of the world was violence.'" Quoted in Robins, op cit.

effusive about how much money could be made dealing the drug, that he urged his king to get the Portuguese back home to start cultivating poppies around the country. He claimed that a shipload could easily be sold every year to India because the "people of India are lost without it."[7]

In the years ahead, opium appeared with increasing frequency in Portuguese travel accounts. In 1516, the ambassador to China, Tomás Pivez de Leira, who was an expert in medicinal herbs, reported seeing opium used regularly in both India and China. Another visitor to India, a Jewish doctor and herbalist named Garcia de Orta, claimed that some people ate up to sixty grams daily to "treat their nerves." He devoted an entire chapter to the drug in a book that was published in Goa, India, in 1563.[8] Similarly, Portuguese doctor and historian Cristoval da Costa witnessed a Hindu scribe downing twenty grams at a time with barely a wobble.[9]

The best source for early opium use in the East, however, is the writer Duarte Barbosa. In his eponymous travel guide, *The Book of Duarte Barbosa* (which is only slightly less florid than Marco Polo's), Barbosa describes coming to the Indian subcontinent in 1501 and going on to live there for fifteen years. One country that particularly caught Barbosa's eye as he traveled around the Indian Ocean was Cambaya on the east coast of Africa, where "luxurious, free livers, [and] great spenders" dwelled. Describing the opium use of Cambaya's King Moordafaa, he wrote:

---

7   Filan, *Power of the Poppy*, 49.
8   Its encyclopedic title was: *Conversations on the Simples, Drugs and Materia Medica of India and also on Some Fruits Found There, in which Some Matters Relevant to Medicine, Practice, and other Matters Good to Know Are Discussed.*
9   Escohotado, *A Brief History of Drugs*, Ch. 9.

He could never give up eating this poison, for if he did so he would die forthwith, as we see by experience of the opium which most of the Moors and Indians eat; if they left off eating it they would die; and if those ate it who had never before eaten it, they too would die.... This opium is cold in the fourth degree; it is the cold part of it that kills. The Moors eat it as a means of provoking lust, and the Indian women take it to kill themselves when they have fallen into any folly, or for any loss of honour, or for despair. They drink it dissolved in a little oil and die in their sleep without perception of death.[10]

Clearly, although regular opium use was a relatively new trend in the East, it wouldn't take long to reach the tipping point in terms of widespread use and addiction, thanks in part to a discovery on the other side of the world.

---

10    Barbosa, Dames, and Magalhães, *The Book of Duarte Barbosa*, 123.

# Chapter 10

# The Two Most Addictive Drugs
# on Earth

While the Portuguese were establishing their violent preeminence in the Pacific, Christopher Columbus was making an equally important "discovery" for Spain across the Atlantic, one that would lead to a whole new way to abuse opium.

Columbus and his men enslaved countless natives, killing them or cutting off their limbs if they didn't mine enough gold.[1] They left behind enough smallpox and other diseases to wipe out huge populations of Native Americans while bringing back enough syphilis to do the same in Europe. But in terms of opium's history their most important and deadly discovery proved to be tobacco.

A few days after first landing on American shores in 1492, while sailing between islands, Columbus came upon a native in a canoe whom he described having "with him a piece of the bread which the natives make, as big as one's fist, a calabash of water, a quantity of reddish earth, pulverized and afterwards kneaded up, and some dried leaves which are in high value among them, for a quantity of it was

---

1    Nunn and Quian, "The Columbian Exchange."

brought to me at San Salvador." Later, a couple of his men who were out exploring the countryside came across "people on the road going home, men and women with a half-burnt weed in their hands, being the herbs they are accustomed to smoke."[2]

Writing years later, a Dominican friar name Bartolomé de las Casas combined Columbus's words and his own recollections to describe what they were smoking as "dried leaves rolled up in the shape of the squibs made by boys at Easter. Lighted at one end, the roll is chewed, and the smoke is inhaled at the other. It has the effect of making them sleepy and almost intoxicated, and in using it they do not feel tired." According to another translator, Las Casas had added that "he knew Spaniards in Española who were accustomed to smoke it, and when their habit was reprehended as a vice, they said they could not leave off," despite that Las Casas did not understand what "pleasure or profit they found in it."[3] [4]

Columbus wasn't clueless about opium itself. He and other explorers including Magellan, Vasco da Gama, and the English explorer John Cabot had all been instructed by their patrons to look for signs of poppy cultivation.[5] But even though they never found any opium

---

2   Columbus and Toscanelli, *The Journal of Christopher Columbus*, 71.
3   While Columbus's words are often quoted, his actual log hasn't survived. Neither has the copy that Queen Isabella ordered to be copied, which is known as the Barcelona copy. Columbus's son Fernando ended up with that one and, in 1538, used it as a basis for a biography of his father, but then *that* copy of the log was lost. At some point, Las Casas, a character in his own right, who knew Columbus and had spent time in the New World, did an abstract that does allegedly include some direct quotes and a whole lot of paraphrasing. There are multiple translations of *his* work also. So, when you read that Columbus said something the only thing you can be sure of is that you can't be sure if he said it or not. He also, by the way, sometimes referred to himself in the third person.
4   Columbus and Toscanelli, *The Journal of Christopher Columbus*, 71n1.
5   John Cabot is credited with being the founder of the tobacco industry in Virginia. And, as we'll see later, Thomas Jefferson was one of the founders of the *opium* industry in Virginia.

poppies growing in America, as soon as sailors obtained tobacco, they quickly began mixing it with opium, thereby combining the two most addictive natural substances on earth into a single smokable blend.

The love affair between tobacco and opium appeared virtually overnight. The apocryphal story of how the two hooked up is that one day an idle Portuguese sailor who had picked up a nicotine habit on his last trip across the Atlantic, and was now on expedition exploring the Spice Islands, decided on a whim to combine it with a little ball of opium. He simply dropped the ball into some sort of pipe or rolled it in a tobacco leaf and lit it up.

Until then, opium had usually been eaten or dissolved in alcohol since it tasted horrid when burned, but mixing it with tobacco made it palatable, and the combination was ideal for sailors. The tobacco helped suppress their hunger, keep them focused, and was even good at treating illnesses like dysentery and malaria. The opium would have relieved the many physical pains that accompanied their arduous lives and relieved the boredom that would inevitably result from months away from home, living on a diet of biscuits, salt pork, and sardines and sleeping out on the open deck while being tossed around by the seas and nibbled on by vermin of all varieties.

There's also a theory that the idea of smoking tobacco and opium together first occurred to a native of Indonesia where they were already smoking a blend of herbs and tobacco and hemp. A similar preparation, called *madak*, that featured opium appeared around the same time along with smoking houses dedicated to its use. It was typically inhaled through a special bamboo flute that would evolve into the long opium pipe seen in pictures from the nineteenth and twentieth centuries. Both release opium's intoxicants by heating rather than burning it, in a way that's similar to modern e-cigarettes and vaporizers.[6]

In any case, opium smoking was not only highly addictive, it was far more potent. Any food or drug taken in solid or liquid form has to go through the digestive system before a portion is absorbed in the bloodstream. Smoking on the other hand sends the drug quickly through the lungs and directly into the bloodstream. There wouldn't be a faster, stronger, or more addictive way to take the drug until the invention of the hypodermic needle in the 1800s.

---

6   Dikötter, Laamann, and Xun, *Narcotic Culture*, 32ff.

# Chapter 11

# The Spice Race

One day, England would become the dominant player in the opium trade—and fight wars to keep it—but only after three centuries of shifting alliances and convoluted geopolitics, during which they fought Spain, the Netherlands, and Portugal for control.

For the first half of the sixteenth century, Portugal was the most powerful European country in the East. In port after port, they landed, planted a flag, built a fort, and negotiated with the local rulers for exclusive settlement and trade rights. After that, the Portuguese king and queen granted trading rights to individual merchants and small trading partnerships to start doing business at that port.

Towards the end of the 1500s, however, a succession crisis in Portugal led to Spain's King Philip II being crowned ruler of both countries and, thereby, given reign over Portugal's extensive holdings in the East.[1] He was known as Philip the Prudent, a rather odd moniker considering his state went bankrupt five times while he was king and, after Queen Elizabeth I refused to marry him, he made

---

1 This unintelligible cross-monarchical folderol started when Henry VIII's Catholic daughter (and heir apparent) Mary died unexpectedly shortly after her marriage to Spain's Philip I, which had joined the two countries in holy matrimony. Elizabeth, who ascended to the throne after Mary's death, wasn't Catholic and wasn't particularly eager to marry Philip, in large part because she would have lost far more of her power and influence to him than vice versa.

a strategically dubious attempt to conquer England, something no country had succeeded in doing since Julius Caesar.

It was, indeed, religion—not wealth or drugs—that led to Spain's downfall. Until then, England had only managed to establish two failed colonies in North America and had virtually no presence in the East. So, when Spain threatened to attack in the late 1580s, England had two choices: sign a treaty that would give them a place at the world trade table or push back.

Philip made the decision easy for England's Queen Elizabeth when he tried to quash an internal rebellion in the modern-day Netherlands, then a Spanish territory. Being under Spain's Catholic religious thumb was fine with the Dutch in the Low Countries (modern-day Belgium and northern France). They were Catholic and religion was easily as important as nationality. But the Reformation had whetted the appetite of Protestants in the north, particularly merchants, to achieve political and economic independence (from both Rome and Spain).

Phillip responded to their rebellion by refusing to allow Dutch merchant ships to land at the ports in Lisbon and Seville to pick up product to distribute throughout northern Europe. Next, he captured the key northern port of Antwerp, sending 100,000 Dutch fleeing further north and cutting off their main supply point. When the Spanish decided to punish the leaders of the uprising—often using techniques they had perfected during the Inquisition—the minor uprising became an all-out revolt.

At that point, the Protestant Elizabeth decided to side with the Dutch, which put Spain and England in a state of war. While their shipbuilders were working overtime to go head-to-head in the English Channel, Elizabeth sent Sir Francis Drake off to launch a guerrilla campaign, attacking Spain itself and invading its trading outposts in the Caribbean and Florida.

Explorer, pirate, slave trader, one-man expeditionary force, and soon-to-be Vice Admiral of the British Navy, Drake had a reputation for being strong on strategy, weak on scruples, but very loyal to his Queen. He'd also been harassing the Spanish since 1568 when he and his cousin John Hawkins were on what they considered an innocent slave-trading expedition to Mexico and their fleet was ambushed and virtually destroyed by the Spanish. They escaped with just two ships, and Drake, who wasn't on either, was forced to save his own life by swimming to shore.

His first act of revenge was attacking the Isthmus of Panama on the Spanish Main, in 1572 and 1573, returning to England with a stolen fortune in silver and gold that made him a national hero. Then between 1577 and 1580, he led the second fleet to circumnavigate the world, attacking Spanish ships at every opportunity and even capturing the Nuestra Señora de la Concepción off the Pacific coast of South America, including its mother lode of gold and silver. From there, he explored the west coast of North America, allegedly claiming large swaths of the California coast for Elizabeth before sailing to Indonesia, where he sweet-talked King Babullah into supporting British traders instead of the Portuguese.

Well aware of Drake's successes, Philip the Prudent became increasingly belligerent. He'd been jilted by the queen, revolted by the Dutch, humiliated by Drake, harassed by pirates closer to home, and pressured by Pope Sixtus V (a troubling oxymoron) to get rid of the insufferable Protestant Queen of England and reestablish Catholicism as the rightful religion of the land.

With Elizabeth's quiet encouragement, Drake got back to work and, by the time Philip had put together a fleet, had struck two Spanish ports, destroying almost forty military and merchant ships.

Within a year Philip was ready to try again. As usual, the infernal

Drake and the rest of the British fleet were more than ready. A series of increasingly embarrassing events followed, culminating in the famous battle in which Spain lost a full third of the ships in their ill-named "Invincible Armada." Scholars often blame Spain's defeat on bad strategy, unpredictable weather, and Drake's particular expertise in setting mothballed ships ablaze and directing them like firebombs into the midst of the Spanish fleet. Still, others have pointed to an apparent burst of enthusiasm on the part of the British sailors, owing to a few inspiring words from the lips of Queen Elizabeth. Before they cast off to defend her honor in the face of the Spanish invasion, she appeared before them on horseback and in full regalia, and inspired them with what would become one of the most celebrated speeches in history:

"I know I have the body but of a weak and feeble woman," she began with false modesty, "but I have the heart and stomach of a king, of England too, and think foul scorn that... any prince of Europe, should dare to invade the borders of my realm; to which rather than any dishonour shall grow by me, I myself will take up arms, I myself will be your general, judge, and rewarder of every one of your virtu'd."

She finished by assuring the troops that "we shall shortly have a famous victory over those enemies of my God, of my kingdom, and of my people."[2]

They did. And in doing so, they paved the way for England to begin establishing its dominant presence in the East.

At the same time, however, the Netherlands, now free from Spanish rule, set its eyes on the same prize. In the next century these former allies would become rivals and combatants in the spice trade, inventing

---

2   Queen Elizabeth I, "Elizabeth's Tilbury Speech."

capitalism, public shareholder companies, and corporate raiding in the process. Together with the faltering Spanish and Portuguese seafaring dynasties, they would end up competing for the land and riches to be found in India and Indonesia, and for access to the biggest market on earth.

# Chapter 12

## The Queen and Her Company

On December 31, 1600, Queen Elizabeth signed with a flourish of her quill pen a document that would lead to the creation of capitalism as we know it, making some people rich, killing countless others, and transforming London into a center of international finance. It would also help establish opium as one of the most valuable commodities on earth.

Like most executive proclamations, hers was wordy and filled with legalese, but unlike most, this one would influence nearly every economic interaction for years to come. It began with a list of 125 men, all described as "well-beloved subjects" of the Queen. Ultimately, 218 merchants signed on, agreeing to invest a total of 68,373 pounds in what Elizabeth referred to as "The Governor and Company of Merchants of London Trading to the East Indies," a mouthful of a name that would end up being shortened to simply "the East India Company."

Elizabeth wasn't too thrilled with signing the merchant's elaborate contract. She'd been working on establishing a fragile truce with Spain and didn't want to antagonize their new king, Philip III.[1] But

---

1  A.k.a., Philip the Pious, who succeeded his father Philip the Prudent in 1598.

after stalling for a year, she ran out of reasons to refuse the merchants' request.

Whether fully aware of the significance or not, she was, for the first time in history, privatizing large-scale international trade by creating a private stock offering that gave a large group of merchants the exclusive right to trade in certain countries—a government-sanctioned monopoly.[2]

This was a *big* deal. Previously, when explorers like Columbus, Magellan, Cortez, or da Gama sailed into the great unknown, the upfront costs came from the royal treasury. After receiving their official charter, the sailors cast off in search of fame and fortune and, if they found the latter, would claim the territory in the name of the monarch and proceed to buy or steal gold, silver, and other commodities from the locals and, in many cases, enslave them and force them back into the mines. When the boats were full they would leave behind a skeleton crew to explore further opportunities and sail back. If they managed to avoid pirates and treacherous seas, they'd return home and pay the royals' investment back with interest.

As Francis Drake had demonstrated, England's other major source of economic power and military advantage came from alliances with pirates who raided ships from other countries. After all, why sail all the way to the East or West Indies when you could wait in friendly, familiar seas and capture fully-loaded ships? While piracy was a dangerous occupation, and pirates' loyalty was always in question, as long as they gave the king and queen their cut, the royals would look the other way.

---

2   To put it in modern terms, this would be as revolutionary an economic game-changer as if the United States, after paving the way with sixty years of space exploration, gave one company exclusive rights to conduct business in space.

While gold and silver were the main currencies, as illegal commerce grew, opium's potency made it extremely useful as an alternate currency since even small quantities were extremely valuable and were relatively easy to conceal from pirates, port managers, and officials looking for their cut.

Queen Elizabeth well understood the potential of the spice trade and opium's unique place in it. Her legal adviser, Sir Francis Bacon, an Elizabethan polymath, studied and wrote about the drug and other narcotics, examining their potential to improve health and longevity—topics that were particularly relevant to the aging queen. Bacon speculated that "if it were possible for young spirits to be put into an old body, it is probable that this great wheel might put the lesser wheels in motion, and turn back the course of nature." Though he strongly approved of opiate use when correctly administered, he also cautioned that while Turks could take opium in large quantities and even before battle, "to us, except in small quantities, and with strong correctives, it is fatal."[3]

Bacon talked a lot about opium's effect on "the spirits," an elusive concept (reminiscent of the Egyptian *met*) that were said to produce all the effects in the body. He believed that it could be used to condense the spirits in order to calm the patient, and he considered it the best palliative in cases involving great pain such as amputations and kidney stones. He also made it clear that opium could excite the sexual passions.

Bacon was also well versed in other narcotics and valuable crops from the East. He referred to the Turks using an herb called "coffee" that they ground up and drank in warm water to invigorate themselves

---

3    All Bacon quotes from "Francis Bacon on Opium, Coffee and Tobacco, Etc."

(although he said that in large quantities it could "disturb the mind"). He also wrote about how a root called "betel" was chewed by the Indians to refresh themselves and for sexual intercourse; and about the increasing use of tobacco, which "affects men with a kind of secret pleasure, so that persons once accustomed to it can scarce leave it off."

In addition, by sourcing opium from the Far East, the British, like the Portuguese a century before, wouldn't be dependent on the "heathens" in the Middle East to meet their domestic demand. Indeed, East India Company ships were given explicit instructions to find and purchase the best-quality Indian opium they could for transport back to England.[4]

The East India Company's first venture involved sending out just four ships in February 1601. They focused on Indonesia, where Drake had managed to make inroads buying pepper, cloves, mace, nutmeg, and opium, and established England's first foreign trade port in Bantam (now Banten) in northwest Indonesia.

The Dutch, however, having pushed out the Portuguese, had every intention of keeping Indonesia for themselves. So, with their new republic's support, several small Dutch companies that had been trading in the East merged two years later to become the "United East India Companies," eventually issuing shares and becoming the first *publicly* traded company. Rather than being limited to subscribers who got to join the club (as in England), anyone could get in on the action—and it was definitely worth getting in on the action. If you invested in the Dutch East India Company at its inception, and held until it was dissolved 200 years later, you would have earned 3,600 percent in dividends.[5]

---

4   "Opium Throughout History."
5   Robins, *The Corporation That Changed the World.*.

The infusion of resources that "going public" gave the Dutch merchants—combined with the country's advanced navigational technology and experience—made it possible, in subsequent years, for the Dutch to continue pushing the Portuguese off the Spice Islands, east and west, north and south, and, as important, to block British attempts to establish any more ports in the region.

Ironically, these failures would end up laying the groundwork for the British East India Company's eventual economic hegemony in the East, in which the company dominated both supply in the opium-producing region of India and demand in the biggest market of all, China.

# Chapter 13

# A 5,000-Year Tradition of Medicine and Moderation

Until the West showed up, China didn't have a problem with opium addiction. Unlike Americans, who took only a few decades to go from medicinal use to widespread addiction, the Chinese used narcotics in their medicinal remedies for thousands of years without getting hooked. Opium was just one of many ingredients in their sophisticated formulas, along with a vast array of flowers, stems, roots, minerals, and ingredients from dozens of life forms—from snake oil and toad secretions to rhinoceros horn and tiger penis. While the first physical records of Chinese medicine only date back to around 500 BC, the characters used (which depict medical terms) date back another several thousand years to the legends of the three founding Chinese emperors.[1]

The first version of those records is an encyclopedic listing of medicinal ingredients that was compiled in approximately 3000 BC. It was written by Emperor Shen Nung, who is also considered the father of Chinese agriculture. According to legend, he tasted hundreds of herbs to determine their medicinal value and, ultimately, recorded about 300 to 400 animal-, plant-, and mineral-based medicines that he considered useful.

---

1    Main sources for this chapter are Huard and Wong, *Chinese Medicine;* Unschuld, *Medicine in China;* and "Classics of Traditional Chinese Medicine."

China's other two legendary emperors played different roles in the development of medicine. Fu Hsi (c. 2000 BC) is credited with writing the *I-Ching* or *Book of Changes*, which was revealed to him in markings on the back of a mythical dragon. Even though it's not a medical text per se, it can give the practitioner a sense of whether his patient's prognosis is good or not. The third founding emperor, Huang Ti, the Yellow Emperor, wrote the *Huang Ti Nei Ching* (*The Canon of Internal Medicine*). Considered the bible of traditional Chinese medicine, it includes a treatise on physiology, anatomy, and acupuncture.

The first definitive record in China of opium being used as a medicine appears in a description in the work of China's first surgeon, Hua T'o (ca. 190 to 265 AD), who made an anesthetic so powerful that he could do abdominal, heart, and spleen surgeries, and even scrape off infected flesh and bone.[2] No matter how harsh the procedure, he claimed, the patient became "as insensible as if he had been...deprived of life"—only to come back to his senses in a few days, convinced he had never felt a thing.[3] The description calls to mind an extremely long-lasting version of popular anesthetics today, particularly Versed, a benzodiazepine that is often used in conjunction with other drugs (including the opiate fentanyl) for procedures as routine as colonoscopies.

While some scholars claim that Hua T'o's soporific brew was based on hemp, according to the famous sinologist Dr. Erich Hauer, Hua T'o's drug of choice was clearly opium.[4]

Not only was his application of opium radical for the time, simply

---

2    Note: Readers familiar with the Chinese art of t'ai chi will appreciate that Hua T'o is also known as the developer of the "Five Animal Frolics," a series of movements often considered preliminary or fundamental to t'ai chi.

3    Simpson, *Anaesthesia, Hospitalism*, 4.

4    Veith, *The Yellow Emperor's Classic of Internal Medicine*, 3.

performing surgery was risky during the height of the Han dynasty because the emperor had made Confucianism the official religion of China and Confucians were appalled at the idea of cutting into the sacred human body. When he had a run-in with one of his patrons—perhaps the opium had worn off too soon—Hua T'o ended up jailed and eventually put to death. He tried to get a jailer to rescue his papers but the man was frightened of being charged himself and so all Hua's secrets went up in flames.[5]

But the execution of Hua T'o did not end opium's use as a medicine in China. When the Song dynasty came into power in 960 AD, there was a resurgence of Chinese interest in the outside world, medical knowledge in particular. In 983, King Qian Chu ordered his doctors to put together a comprehensive revision of the *Pên tsäo* (the encyclopedia of Chinese medicines). In this version, opium was called *ying-tsu-su* (*su* being the word for a "pod") and it was recommended as a cure for dysentery. There's no indication at all that the drug was abused or led to addiction.

The fact that a theme of balance and restraint is common in Chinese belief systems also played a role in preventing opioid abuse in the country despite its long tradition of medical use. From Confucius, who approved of alcohol but disapproved of drunkenness, to Buddha, who believed that people should refrain from intoxicants altogether, to Lao Tzu, who cautioned, "The flame that burns twice as bright burns half as long," there are many teachings that promote moderation.

Whether it was owing to a combination of these cultural influences, the discriminating use of the drug by physicians themselves, or even

---

5    The one exception was his method of castration, which, for better or worse (depending on which side of the blade you're on) continued to be practiced for many years.

the weaker variety of opium being cultivated in China, all evidence suggests that, in the millennia before the European merchants arrived, Chinese opium use was limited, well supervised, and safe, belying the racial stereotypes that the West would soon foist upon the Chinese as being, by nature, incorrigible addicts committed to drawing the rest of the world into their sordid drug culture.

If anything, it was the opposite.

# Chapter 14

# Opening the China Market

Long before opium became an important part of the China trade, the factors that would make conflict with the West inevitable were firmly in place—in particular, the country's inflexible bureaucracy and closed-door policy toward trade.

From the opening of the Silk Road, merchants caravanning overland to China maintained a small trading network. A few Arab ships did explore the South China coast in search of a safe harbor. In fact, Arab merchants looted and burned the pivotal harbor of Canton in 758, but even in the wake of that temporary victory, trade with the West remained severely restricted.

A few hundred years later, however, the Song dynasty (960–1279) opened the door a crack when the Emperor Taizong sent eight court eunuchs on missions to invite traders to come to China.[1] Lest this new welcoming attitude be considered a sign of weakness, however, Taizong's successor, Emperor Zhenzong, made an official proclamation that the state would maintain a strict monopoly over all

---

1 Eunuchs, of course, have a notoriously bad rap, potently speaking. They were trusted (and usually trustworthy) allies of the Emperor who could be assured they wouldn't seduce—or at least not impregnate—his wife or daughters and, in the process, father any potential rivals for the throne. Several eunuchs who served during the Song Dynasty were also famous military commanders and explorers. See Cartwright, "Eunuchs in Ancient China."

commerce, which meant everyone who wanted to trade in China had to pay the price. The Chinese weren't interested in developing close alliances with Western countries, they were interested in trade, and on their terms.[2] So, in the years to come, while merchants might be allowed to build an office or even a small warehouse, they weren't allowed to establish the kind of permanent, more or less self-governing settlement in China's commercial centers the way they would in other countries.

In part, this protectionism was based on China's conviction that their culture was superior to the West's and they didn't want it diluted by foreign influences. Equally important was the fact that China simply didn't need to open its doors in any major way because it was a massive country, rich in its own natural resources, and had a population that could grow or manufacture virtually anything it wanted.

After describing the abundance of China's natural resources, skills of its people, and vast *internal* trading network, the eighteenth-century Dominican missionary Gaspar da Cruz concluded: "That which showeth much the nobleness of the country, the plenty and riches thereof, is, that all these ships bringing great store of merchandise of cloths, silks, provisions and other goods, some do go into the land, others come from within the land, and nothing cometh from without China, neither goeth of it."[3]

\* \* \*

Eventually, China did allow significant sea trade, but limited it to the 1,500-mile-long, 70-mile-wide Pearl River estuary in the middle of

2   Schottenhammer, "The 'China Seas' in World History."
3   Da Cruz, "Treatise in Which the Things of China Are Related."

Guangdong, the province that would soon be the center of the opium trade and which remains a major Chinese commercial hub.

In either 1513 or 1514, Portuguese explorer Jorge Álvarez, the first Westerner to reach China by sea, planted a flag on an island at the mouth of the Pearl River and "claimed it" for the Portuguese king, Manuel I ("The Fortunate").[4] The Ming emperors responded with understandable hostility and, after a few skirmishes, Emperor Jiajing agreed in 1557 to allow the Portuguese to dock their ships at the harbor of Macau, an island just off the coast at the mouth of the estuary. Since it was remote enough to prevent foreigners from infiltrating the mainland, but close enough for them to do business, the emperor granted Portugal the right to virtually govern Macau, albeit under close Chinese scrutiny.[5]

The Dutch spent the next seventy years trying to pry Macau away from Portugal, finally giving up only after a bloody battle in 1622, a humiliating loss that they implied was "unfair" because it was primarily at the hands of African slaves brought over by the Portuguese. Remarkably, Portugal de facto ruled Macau until 1887, when they actually made it a colony, only to give it back to China in 1999. Rather than an indication of weakness, this arrangement was based on China's pragmatic appreciation for how difficult it was to administer their remote islands effectively; it preferred to leave that to a relatively reliable trading partner.

A few decades later, however, Emperor K'ang-hsi decided that Canton, at the source of the Pearl River estuary, where it meets the mainland, was the best place for the Chinese to monitor trade

---

4   Most likely Lintin, which would become an important commercial island.
5   Unless indicated, the source for most details in this chapter is Van Dyke, *The Canton Trade*, an excruciatingly detailed but remarkably readable book about the trade in Canton from 1700–1845.

with foreigners, so he insisted that imports go no further than there. Canton's location may have made it somewhat easier to manage than Macau, but it was still 1,500 miles from Beijing and a world apart. Even if the emperor was getting accurate information and sending clear instructions, imperial edicts could arrive weeks after the fact, and depending on who was passing those instructions and what money was being exchanged along the way, the words could be subject to extensive editing, or never even arrive.

Canton was far away culturally as well as geographically. In terms of rules and regulations, their attitude was, "The mountains are high and the emperor is far away." For their part, the Ming emperors in Beijing "regarded their subjects to the south—the Cantonese—as no less than witches and sorcerers, their language unintelligible and their culinary predilections downright disgusting. It was therefore fitting that they left it to the Cantonese to do business with the even more outrageous barbarian traders."[6] In other words, the emperors were willing to let what happened in Canton stay in Canton unless the gold the foreigners were bringing to buy goods stopped flowing into the treasury.

Still, the emperor couldn't ignore the West's growing influence in Canton indefinitely, and the difficulty of maintaining control from so far away would prove to be yet another weakness in China's ability to respond to British aggression.

\* \* \*

As we've said, opium *abuse* actually came a little late to China. At first, the government was more concerned about tobacco, whose use was

---

6   Lonely Planet, "History of Hong Kong."

spreading like wildfire, especially along the coast. Soon, Portuguese and Dutch sailors began introducing the doubly addictive habit of mixing little balls of opium into their tobacco, a mixture known as *doop* (yes, "dope"). Raw opium is a sticky goo that doesn't burn easily and, when it does, makes the user choke.[7] As we've seen, the burning tobacco vaporizes the morphine and other alkaloids in the morphine, which is a far more effective way for the body to metabolize them. The practice seemed innocent enough to Chinese dockworkers and merchants, but neither they nor the emperor seemed to understand that the Middle Eastern opium was much stronger than the mild homegrown variety the Chinese doctors formed into pills to treat diseases.[8]

At first, smoking these pills in tobacco seemed to just make them more palatable. Soon, however, it became clear that it also made the opium far more addictive.

In 1638, the Ming emperor Chongzhen, concerned and outraged, made tobacco use a crime punishable by decapitation. After the edict, most people just started doing their tobacco smoking in secret. Next, some simply figured out a way to smoke their opium straight by filling a clay bowl, attaching it to a long tube with a tiny hole, and using an oil lamp to heat it until it vaporized but before it burned.[9]

A few years after Emperor Chongzhen's tobacco edict, the Manchus overran Beijing, which marked the end of the Ming dynasty and

---

7   For details, see Martin, *Opium Fiend*, 66.

8   By then opium "pills" were a common treatment for various diseases, particularly dysentery. (They referred to them as "pen yen," which explains the derivation of the phrase having a "yen" for something.)

9   These primitive opium pipes evolved into a remarkable Chinese craft: objects of exquisite form and ever-improving function. However we might feel about what the pipes were used for, from a solely craftsmanship perspective, the destruction of thousands of them during various government crackdowns in years to come is a great loss. See Martin, *Opium Fiend*.

beginning of the Qing dynasties. Ethnically different, the Manchus established new cultural norms, including that women were no longer to bind their feet and men were to shave their forehead and tie their hair in a braid in back—a style known as a *queue*. In addition, shortly after coming to power, Fulin, the Shunzhi emperor, rescinded the tobacco edict. But it was too late—not just for the people who had been decapitated but also for those who had switched to straight opium. They not only continued to smoke at home, but also began experimenting with the latest breakthrough in interior design—primitive opium dens.[10]

The plant extract that people had been using medicinally for centuries became a regular habit for some Chinese, an addictive one for many, and the source of much pain and suffering throughout the country for generations.

---

10   Filan, *Power of the Poppy*, 239.

# Chapter 15

## Great Britain "Invades" China

Even though "India" was part of the East India Company's name, at first it had largely overlooked that country's potential in favor of trying to get in on the action in the Spice Islands. Slowly, however, thwarted by Spanish, Portuguese, and Dutch control of key areas such as the Philippines and Indonesia, English merchants began to focus more on the interior of India.

By the seventeenth century, most of India was part of the Mughal Empire. While the British had strategically established several small outposts on the coast, the Mughal emperors continued to give Portugal preferential status in terms of the ports they were permitted to use and the price they paid for product. Then, in 1612, a small but important battle between four of the East India Company's armed galleons and a larger Portuguese force took place. Although a standoff, the battle made it clear to the Mughal emperor, Jahangir, that the Company was a force to be reckoned with.

Over the next decade, he gave the British merchants increasingly favorable terms, and they began exporting large shipments of Indian cloth from the port city of Surat back to eager customers in Europe. Jahangir had one other reason to collaborate with the British: unlike the Dutch and Portuguese, they were there to do business, not conquer or convert. As England's ambassador Sir Thomas Roe put it, "If you will profit, seek it at sea and in quiet trade."[1] (Granted, Jahangir may also have been influenced by the habit Portuguese commanders had of tormenting Mughal Muslims and blocking their pilgrimages to Mecca.)[2]

For the upper classes and skilled artisans, the Mughal Empire was an enlightened and opulent civilization tolerant of different religions, with a well-established network of loyal administrators who kept watch over the nominally independent local chieftains. At the bottom rung of the economic ladder, however, were millions of over-worked peasant laborers who wove textiles, manufactured silk, mined gold and silver, carved diamonds, and crafted metals under the direction of those artisans. They also cultivated spices, sugar, indigo, and opium, under the direction of the chieftains.

The empire was also the home of remarkable architects, including Emperor Shah Jahan, who succeeded Jahangir and is credited with building the Taj Mahal. A committed opium user, he drank his opium with his wine and decorated his wife's tomb with poppies,[3] an indication of how the drug had infiltrated daily life—at least for the upper classes.[4]

As their people increasingly began to use the drug, however, the Mughal leaders saw no reason to rely on imports since their own climate was perfectly suited for growing poppies. Indeed, they soon became a major producer and then exporter. Indian opium didn't have as much morphine content as the Turkish variety, but transportation costs were so minimal compared to shipping it from the Middle East that dealers and users could always make up in quantity what was lost in quality.

---

1   Robins, *The Corporation That Changed the World*, 46.
2   "Battle of Swally Hole | Indian History."
3   Hayes, "Opium Wars in China."
4   While strictly regulated, opium remains an ingredient in Indian Ayurvedic medicine to this day. See Tiwari, "Grocer Sells Ayurveda Medicines with Opium."

As cultivation took hold, the British, by then very familiar with opium cultivation and processing from their alliances and trade with Turkey, made themselves increasingly useful as consultants and contractors to growers and processors. Meanwhile, they waited for signs that the Mughal Empire might crumble, which they considered inevitable. Once no strong centralized government was left to push back, they could take control of the production itself and, essentially, the people producing it.

And so it happened. Different regions, particularly in the south, started fighting the Mughals for territory and independence. That, as well as infighting between tribes, sparked a multifront war that dissolved the empire. The British worked on the sidelines, offering money and weapons to groups that at various times appeared to be useful, malleable allies.

Some leaders, however, weren't willing to cave under British pressure. Most famously, in the province of Bengal, a local nabob (governor) in Calcutta named Siraj ud-Daulah was so annoyed by the liberties the British were taking, he destroyed their warehouse and threw some of its officers into the infamous "Black Hole of Calcutta," so named because 123 of the 146 men who spent the night there died of suffocation and heat exhaustion.

The British—as they would soon demonstrate to the Chinese as well—took umbrage at the idea of being unceremoniously evicted, even if it was from a region where they had never been fully welcome in the first place. Robert Clive, the indomitable British governor of Bengal, later accused of being an opium addict and "unstable sociopath," prepared to take revenge.[5]

---

5   Dalrymple, "The East India Company."

Clive would be remembered for his rampant corruption but also given a great deal of the credit for the East India Company's success in India. He successfully commanded the Company's armies and defeated Siraj ud-Daulah's forces in the town of Plassey, which gave the Company virtual control of Bengal and Bihar. These two major opium-growing regions had the added advantage of having borders on the Bay of Bengal to the south and China to the north.

Before long, the British managed to turn these key opium-growing provinces into quasi-sovereign, quasi-colonial regions, which gave them two important advantages over the long-established Middle Eastern sources: they didn't need to buy the opium they wanted to sell and had no financial obligation to local rulers.

The means of production were firmly in British hands, and a huge captive market that was gradually getting addicted was just over the border. The only things that could stop them were the faint moral misgivings back home, where some government officials had begun to question the ethics of selling such an addictive drug to people in a country whose government was trying to crack down on drug abuse and maintain at least a semblance of its closed-door policy.

\* \* \*

At dawn on August 17, 1780, Philip Francis, a member of Bengal's five-man Supreme Council, took seven steps, turned around, and prepared to shoot his fellow council member and bitter enemy Warren Hastings. But his powder was damp—his gun didn't fire. Hastings politely waited until Francis reloaded and the two experienced officials, if inexperienced duelers, gave it another try. This time, both guns fired. Francis missed. Hastings didn't. Although he'd only been hit in the shoulder, Francis fell to the ground and cried he was dying. That would prove to be his third error in just a few minutes: he lived. But he went back to England and Hastings went back to walking the moral and political tightrope between his appreciation for the native culture and his government's schizoid approach to the morality of the opium trade.[6]

Hastings had first come to India at eighteen, serving as a clerk for the East India Company. He quickly developed an uncommon appreciation for Indian culture, religion, and tradition. First, he learned the native languages, Urdu and Persian, which made it possible for him to serve as an ex officio mediator/ambassador who proved successful at finessing the relationship between the British government and India's shifting patchwork of administrative fiefdoms. After the kind of chaotic adventures and misadventures that went with the territory, he ended up volunteering to help Robert Clive avenge the Black Hole killings in Calcutta. Clive saw young Hastings's potential and soon put him to work increasing Britain's power in the region by finding ways to manage and manipulate the Bengalese leaders.

Hastings, however, liked and respected most of those leaders. And he didn't like or respect the way his country was treating them. Eventually, he became so troubled by the situation that he went home, where he lived large and went deeply into debt. At that point, with his fortunes depleted, Hastings reluctantly went back to India, where he began to make a great deal of money again working for the Company.

After a series of internal conflicts and famine in India had slashed the Company's stock price, he was appointed the first governor-general of India and told to see if he could get things on a more secure financial footing. He held the position from 1773 to 1785, during which he had to juggle an impossibly delicate position: simultaneously representing

---

6  See Willasey-Wilsey, "The Enigmatic Warren Hastings"; and Robins, *The Corporation That Changed the World.*

a private company that was driven by profits, and a deeply divided government that was finding it increasingly difficult to reconcile its ethical and financial priorities.

However morally indefensible his company and kingdom, Hastings did the best he could to behave honorably in the service of both. He accepted the fact that the Company had literal boatloads of Indian opium they had to sell, and the potential profits could trump any crisis of conscience back home. All Hastings could do is keep the forces of money and morality from bringing business to a standstill by managing a procedure that may have looked respectable but didn't fool him or anyone, except the many merchants and politicians who wanted to be fooled. Private traders from various countries would buy the East India Company's opium and sell it anywhere they wanted through a network of dealers and smugglers. However, it was an open secret that these private traders could *only* buy Indian opium directly from the Company and that most of it ended up in China at the Canton ware-. houses of increasingly powerful trading houses who were able to buy the opium from ships flying any old flag... including Jolly Roger's.

What was the alternative? The East India Company needed the gold and silver they were paid for their opium to buy more tea and other products that were prized back home. And, since the Company was selling opium to middlemen traders to whom they could legally sell it, the principals could convince themselves, if nobody else, that they were adhering to the rules of law. As one Company director wrote, "Whatever opium might be in demand by the Chinese, the quantity would readily find its way thither without the Company being exposed to the disgrace of being engaged in an illicit traffic."[7]

---

7    Blakeslee, *China and the Far East*, 163.

Eventually, however, Hastings decided to put an end to the charade by quietly sending two Company ships full of opium directly to China.

Although the directors of the Company were undoubtedly pleased with the increased profits that came from cutting out so many middlemen, they were still worried about being tied to the illicit trade. It turned out they had good reason because this was the last straw for a group of British politicians who believed that the opium trade was a disgrace. To protect itself from government censure, the Company had to maintain its deniability. And Hastings was the perfect scapegoat.

The philosopher, politician, and moralist Edmund Burke led the charge to impeach him, accusing him of a variety of high crimes and misdemeanors. Burke claimed that in charging Hastings, he was acting on behalf of the people of India (which would have been a surprise to them), as well as "eternal laws of justice and human nature itself, which he has cruelly outraged, injured, and oppressed, in both sexes, in every age, rank, situation and condition of life." The impeachment trials lasted almost a decade before Hastings was declared innocent of all charges.

As with virtually everything connected with the opium trade, nobody was in a position to cast the first stone. Even the insufferable moralist Burke—whose litany of charges allegedly took two full days to read—was clearly guilty of cruel and unusual hyperbole, being unforgivably verbose, and behaving shamelessly holier-than-thou.[8]

If anyone emerged with a relatively clean karmic slate from this era of the China trade, it was Hastings. While on a day-to-day basis he had to make some less than admirable decisions, over his years in India, Hastings developed a great respect for Hinduism and did his

---

8    Burke, "Edmund Burke Quotations."

best to keep his countrymen from ignoring or outright insulting the local culture and traditions. A generous philanthropist when he could afford to be, he sponsored the first English translation of the Bhagavad Gita and supported the building of a Buddhist temple and a school for Muslim students. A century later, Nehru credited him for helping to keep India's heritage alive during the colonial period.

Still, if we can learn anything from Hastings's story it's that when it comes to opioids—whether buying, selling, or railing against them— it's hard to find a moral high ground.

# Chapter 16

# Trading Opium in Canton: "The Complicated Machinery of Evasion"

By the late 1700s, harvesting opium from the fields of India and exporting it to China was a complex process involving multiple traders, sailors, and government officials playing multiple roles in multiple plots and subplots. When a ship finally began its journey up the Pearl River, the logistics became even more complicated, especially if there was contraband on board.

The most important ports on the estuary were Macau, which was located at the mouth, and the city of Canton, which was inland and the only port open to foreigners. At Macau, a ship paid its first tax, received its first permits, and was assigned a local river pilot who knew the estuary well enough to guide the ship between customhouses. At each customhouse on the way upriver, there were soldiers, tollhouse keepers, and inspectors.

While China could keep a close eye on the comings and goings of big ships at the customhouses, the rest of the estuary was a smuggler's and pirate's dream and a law enforcement officer's nightmare—a maze of islands, inlets, and shallow water that offered plenty of places to unload contraband into smaller, lighter boats that could deliver it to local villages along the river where residents would buy from smugglers and even harbor those who needed places to hide out. As one observer noted:

No wonder that smuggling in every form has been long carried out to such a notorious extent.... The communications by water from one point to another, and in the interior of the country, are so numerous, and so interwoven with each other, that it would be impossible for any system of fiscal regulation which the Chinese could adopt, to act efficiently against the complicated machinery of evasion which so easily [can] be put in operation.[1]

The only thing standing in the way of rampant corruption were officials known as "tidewaiters," who commanded small boats that were chained to mercantile ships as soon as they entered the estuary. Their job was to monitor the ship in transit and prevent illegal trade by regularly counting crews, commodities, and cannons—ensuring their numbers matched the paperwork they had received from Macau. While the name sounds pedestrian, tidewaiters were high-ranking officials who lived in their own comfortable quarters using the services of writers to keep the records and servants to maintain their floating offices. Most of them did work hard. There were regular crackdowns on any unexplained product "shrinkage," and efforts to identify corrupt officials at the tollhouses. Still, even though tidewaiters were more difficult to corrupt, like everyone else, they had a price. Eventually, payoffs to facilitate smuggling became as routine as paying the official taxes on *legal* imports. As long as the costs of doing business were predictable, foreign ships sailed confidently upriver, knowing they had inoculated themselves against every force empowered to oppose them—except perhaps for the pirates, who were often their partners in crime.

Once a ship made it to the island of Whampoa—sixty miles

---

1   Hall and Bernard, *The Nemesis in China*, 124.

upriver from Macau but still twelve miles from Canton—additional regulations slowed the process to a crawl. There were three toll-houses on that island alone, and only after those tolls were paid could negotiations begin with one of the thirteen powerful hong merchants who controlled all trade in Canton.

Not only did the Chinese not want foreign ships to go any farther, but by then the water was so shallow that only smaller sampans and junks commissioned by the foreign merchants could ferry product to the warehouses in Canton proper and return with cargo for shipment back to Europe.[2] Many ships at Whampoa became floating ware-houses, storing inventories until the price was ripe for a profitable trade while, at the same time, buying up Chinese exports (especially tea) from the smaller boats shuttling back and forth.

While paintings of all these sampans, sailboats, and schooners show scenes of majesty and tranquility, Canton rapidly grew into an extraordinary commercial center, teeming with exotic merchandise and the sounds of languages from around the world. There may have only been a few hundred large foreign ships a year arriving at Whampoa by the early 1800s, but the Chinese had thousands of vessels of all sizes working the Pearl River and the South China Sea, and Western companies and countries had hundreds more.

In the heyday of the opium trade, smugglers would rendezvous with foreign traders at quiet coves in Whampoa, often at night, but sometimes even in broad daylight since both parties knew that the likelihood of enforcement was minimal.

The final piece of logistics required to avoid additional tariffs or

---

2   The fact they are called "junks" has nothing to do with their quality. Some were as beautiful as full-sailed schooners. The origin of the term is etymological…linguistic…although there is some dispute over which ancient language contributed to this particular Western misconception. See Harper, "Junk."

confiscation, before the ship departed from Whampoa, was to make sure it left with Chinese goods of similar value and weight as what it arrived with—in other words, an appropriate cargo load for a ship that size. Otherwise it would raise suspicions that the boat had come primarily to sell contraband.

The best way to get opium to the mainland without being caught was to avoid having to ship it up the estuary in the first place. This was done by using the "chit" method (similar to an IOU). A ship arriving at the mouth of the estuary would transfer its contraband to a "storage ship" before continuing upriver with its legal cargo. Meanwhile, smugglers would have paid hard currency to unscrupulous merchants in Canton to buy chits. Back down at the storage ships, they could use these chits to buy the opium from the merchants' associates. Only the smugglers were really taking any risk and, since they were charging "retail prices" for the opium they sold to local dealers, it was well worth it.[3]

Another strategy that was used to convince officials that the ship hadn't offloaded any opium took advantage of the fact that the Chinese were eager to stockpile rice in the eighteenth century, owing to the horrible famines they had experienced in the wake of crop failures and the country's ballooning population. Starting in the mid-1700s, the emperor began lowering import duties on rice to encourage foreign boats to transport it from other countries in the Far East. A boat that was 700 tons or more—the size of most European ships—would get a 50 percent discount, and ones carrying about half that much (the size of most American vessels) would be discounted 30 percent.[4] It

---

3   Downs, "American Merchants and Opium Trade."
4   The Chinese used a unit of weight called a "picul," which, like a chest of tea or opium, weighed between 130 and 140 pounds. The "tonnage" here is based on the "picul" as described by Van Dyke, *The Canton Trade*, 135.

was therefore very advantageous to bring a large ship full of profitable rice to Whampoa—particularly if its state rooms, storage rooms, cabin, and every other possible nook and cranny were stuffed with opium, because a ship that had arrived loaded with heavy rice wouldn't be expected to weigh as much when it departed—especially considering the rudimentary weighing technology available.

By this time, merchants from Britain, Portugal, the Netherlands, France, and other countries lived, worked, and stored their merchandise in a reverse gated community that kept them from entering the city. It became known as the Canton Colony, a place where Europeans could pretend they were still living in Europe and the Chinese could, for the most part, pretend they didn't exist.

These merchants were only allowed to do business with a dozen families, the hongs, who controlled Canton's trade. Hongs could become extremely wealthy from foreign trade. Howqua, the chief hong merchant, was at one point worth the rough equivalent of $5 million today. Still, some went bankrupt, either because prices fluctuated wildly and they didn't buy or sell at the right time or because foreign merchants found ways to double-cross them.

The hongs were controlled by *Hoppos*, government officials who functioned like customs superintendents. They were supposed to oversee all commerce in Canton, but were readily corruptible, too. A Hoppo's term lasted for three years and was unpaid—unpaid by the government at least. They were permitted to charge merchants, however, 10 percent of every shipment plus another 2 percent to maintain their staffs. But most Hoppos needed more than that to support their opulent lifestyles, so they demanded additional payments from any hong merchant who wanted to stay in the game. The whole system became one of payoffs and willful ignorance.

The other key player in the Pearl River's international trade was the British governor-general. He was responsible for negotiations with China on behalf of the English government and British merchants (arrangements with which merchants from the other countries always complied). The Westerners were at a significant disadvantage because, at first, the emperor insisted that all of these negotiations take place in Chinese so his merchants could maintain complete control. To make sure the Westerners couldn't question the translation, he made it a capital offense for any of his people to teach them Chinese. With time, however, Chinese traders, river pilots, and customhouse overseers all adopted pidgin English, and used it whenever possible to conduct everyday business. This prototypical "pidgin" language consisted mostly of English, with notes of Cantonese and other Chinese dialects as well as words picked up previously from Portuguese and other traders. Its legacy lives on in vernacular English today in phrases such as "long time no see," "look see," and "no can do," which all originated in Canton-based pidgin English.[5]

One of the hong's most important jobs was to tell the Hoppo when it was time to get down for the measuring ceremony, since dealers couldn't board merchant ships to view what they were buying until that ceremony began.

The phrase "measuring ceremony" doesn't do the event justice. It was a celebration, complete with choreographed cannon salutes, endless eating and drinking, speeches of excruciating formality, talent performances, and the exchange of gifts, which primarily involved the foreigners ingratiating themselves to the Hoppo by offering him European goods like mirrors, jewelry, or music boxes at well below cost.

---

5   It had nothing to do with pigeons and little to do with standard English or Chinese. It was, however, a transliteration of the Chinese word "business" or phrase "pay money." See Wei, "Chinese Pidgin English."

As trade grew, there were soon too many foreign ships arriving in Canton to do this for each one, so the Hoppos started measuring multiple ships in one ceremony. Eventually, only some of the ships got the full treatment.

As the trading infrastructure in China matured, a financial bond business also began to take root, as insurers discovered that they could sell "junk bonds" to the merchants so they could hedge against the not insignificant possibility that their ships would catch fire, be fired upon, be attacked by pirates, or sink of natural causes. (These insurance bonds usually paid investors around 40 percent per voyage because of the scope and extent of the risk—if the ship ended up at the bottom of the ocean, neither the vessel itself nor the goods aboard could even be sold for salvage.) There were also significant lending opportunities for risk-averse investors who could typically make 10 percent or so lending to merchants who needed money to buy product they wouldn't be able to sell for a while.

Landlords and developers profited as well, building or buying structures in the foreign zone of Canton, and renting them out to merchants for housing, offices, or product storage. In addition to collecting rents, some landlords offered, for a price, to protect the Westerners' warehouses from government officials coming to inspect them at inopportune times.

Finally, there were "compradors," the ultimate middlemen, who were responsible for providing the thousands of major and minor foreign players in the Pearl River's trade network with food and supplies—everything from beef, capons, eggs, fish, goats, and oranges to caulk, lamp oil, paint, and other basic boat supplies. They even "supplied" merchants with short-term loans to carry them over between shipments, or while they waited for prices to go higher. While the compradors' business arrangements with Westerners were complex, they were also quite lucrative. In fact, the comprador system led to the creation of several large corporations that accrued so much wealth and power from the heyday of Canton trade that, until recently, they continued to serve as middlemen between East and West.

Throughout all the vicissitudes of the opium trade and upheavals in Chinese politics that have taken place over the last 400 years, many of the ways that China and the West conduct business today can be traced back to the Canton trade.[6]

\* \* \*

As the byzantine logistics of the opium trade evolved in Canton, where merchants and clever smugglers regularly proved they could subvert just about any regulation, a succession of emperors 1,300 miles away in Beijing addressed the situation as anything from a minor annoyance to a full-blown crisis.

Their battle against the illegal drug trade had begun back in 1638 when Ming emperor Chongzhen—thinking that tobacco, not opium, was the real problem—made its use or distribution a crime punishable by death. Several decades later, as the epidemic of opium use—particularly along the coast—became increasingly hard to ignore, Emperor Yongzheng prohibited the operation of smoking houses and the possession of opium for anything besides medicinal use. He ordered that opium smugglers and owners of opium dens be strangled and threatened to confiscate any ships carrying any opium. The fact that, at one point, the emperor was willing to cut off the heads of people for simply smoking tobacco shows that, for the Chinese, no option was off the table.

---

6   Vines, "The Decline of Compradors."

By the 1730s, however, it was clear even those imperial edicts were having little effect—up to fifteen metric tons of opium were arriving every year from India's Bengal province alone. So, in 1757, the Qianlong Emperor established what were known as the "Eight Regulations." They were designed to govern not just the way business was conducted, but also the behavior of all Westerners and the Chinese involved in it:

1. No warships were allowed to enter the Pearl River.
2. Europeans were only allowed to live in Canton during the actual shipping season (September through March) and were not permitted to bring wives or weapons.
3. All the pilots, boatmen, and agents working for foreigners had to be licensed.
4. Only a fixed number of servants could work for the foreigners.
5. The use of sedan chairs and boating for pleasure was forbidden, along with excursions into Canton itself. Guided visits to the public gardens on Honan Island were allowed for groups of less than ten as long as they returned before dark and did not drink liquor.
6. All business had to be carried out through the monopoly guild of local merchants, known as the cohong.
7. No smuggling and no credit were allowed, and the cohong had to file a declaration that no opium was on board.
8. All ships of any size coming to trade *must* anchor at Whampoa, where loading and unloading would proceed under imperial inspection.

While seemingly strict and all-encompassing, these regulations had little effect. Even when additional rules were put in place during the late 1700s—including laws that subjected consumers and dealers to severe beatings or even the death penalty—the opium trade kept growing. By 1800, annual imports from Bengal alone climbed to as much as 300 metric tons, twenty times what they'd been when the Eight Regulations had been put into effect. And by 1838, at least 2,500 metric tons of opium were arriving in Canton every year.[7]

The numbers imply that everything the successive Chinese emperors did to manage the opium trade always proved to be too little, too late. The regulations usually either weren't enforced, weren't getting to the root of the problem, or weren't enforceable in the first place— not only because of the cleverness of those intent on breaking them, but because, eventually, opium became so widespread that it was used by the many clerks, secretaries, and other bureaucrats who were allegedly faithfully executing the emperor's orders. While addiction was indeed an issue in seventeenth- and eighteenth-century China, overdose deaths remained extraordinarily rare.

When, however, the emperor's own son died of an overdose during the 1830s, it was the last straw.[8] Something *had* to be done. So he sent one of the most powerful law-and-order vigilantes of all time to put a stop to opium imports—legal and illegal—once and for all.

---

7    The measure of imports was always an approximation. Typically, it was based on a standard "chest" of Indian opium. A "chest" was a box made out of mango wood (which, while plentiful, was subject to fungal or insect attack). Each chest held twenty large opium balls and weighed from 120 to 150 pounds. While the numbers didn't always add up—especially when comparing the amount that left India and what arrived in Canton—every analysis of data from the period shows that the trade kept multiplying.

8    One writer claims all three of his oldest sons died and their source for opium was the son of the governor of Canton. See Inglis, *Forbidden Game*, Ch. 6. Another says that the emperor was so angry at one of his drug-addict sons, he struck him dead in one blow. See GlobalSecurity.Org, "Manchu Emperor Daoguang." Regardless, the problem had come too close to home to ignore any longer.

# PART IV

---

## The Opium Wars

# Chapter 17

# Two Letters that Could Have Prevented a War

Two letters that bracket the two-century buildup to the Opium Wars never arrived, at least in what could be called a timely manner. Either one could have changed the course of the history of opium and English-Chinese relations.

The first was sent in the spring of 1602 by Queen Elizabeth I of England, who wanted China's Wanli Emperor to give preferential terms to the new East India Company. However, the letter went missing until it was discovered in the grain bin of a Lancashire farm in 1984 and ceremoniously delivered to China, 383 years late.[1] At the time, in spite of its victory over the Spanish Armada, England was still the second-largest European power behind the Spain-Portugal alliance and, when it came to global trade, a *distant* second.

The Spanish-Portuguese alliance controlled many of the islands in the Far East as well as the two existing routes to China—south around the tip of Africa or southwest around the tip of South America. If there were a northern route, England would have their own direct access to the Chinese market. So, Elizabeth gave her letter of introduction

---

1 France, "China Receives Letter from Queen Elizabeth 1—383 Years Too Late."

to an explorer named George Waymouth and sent him off to find the much-sought-after Northwest Passage.

Much of her entreaty does read like a letter of introduction— one royal to another—asking in that arrogantly humble way of kings and queens that a loyal subject be well treated. But, in retrospect, the letter was the first salvo in the formation of a potentially earth-shattering alliance that never was, and its contents demonstrate just how unfamiliar with each other the two cultures were at the time. For instance, Elizabeth includes translations in Latin, Spanish, and Italian, because she did not know which Western languages were currently in the emperor's favor or what translators he had in his court. Global economics aside, the illuminated parchment she wrote her letter on is stunningly beautiful. The Chinese may have considered the British to be barbarians, but the most imperious calligrapher would have to be impressed by her signature alone. Still, the letter was formal to the point of obsequiousness, something totally out of character for the imperious Elizabeth.

"We have received divers and sundry reports both by our own subjects and others, who have visited some parts of your Majesty's empire. They have told us of your greatness and your kind usage of strangers, who come to your kingdom with merchandise to trade." She goes on to describe her hope to "find a shorter route by sea from us to your country than the usual course that involves encompassing the greatest part of the world." She acknowledges that similar expeditions had not made it, "presumably because of frozen seas and intolerable cold." Clearly, she was ready to sacrifice the life of yet another of her adventurous loyal subjects to prove her willingness to go the extra nautical mile for the sake of a mutually profitable direct relationship with China.

But the perilous journey was Waymouth's problem. Elizabeth

focused on composing her letter, in which she tried to ingratiate her-self to the emperor by apologizing for not sending along more gifts, adding that she had, however, included a wide variety of sample products, that there were a lot more where they came from, and that her official sales rep, Waymouth, could answer any questions.

Waymouth would never get the chance to deliver the note. After much of his crew got sick and the rest staged a near-mutiny near the Hudson Strait, he gave in to their demands and turned back, arriving home in September. (It was just as well. His boats would have never made it through the only possible northwest passage, which wasn't found until Roald Amundsen did it in 1905.)

One can imagine him kneeling down before Elizabeth, bowing his head, and handing the letter to her, prepared to spend the rest of his days in a dungeon somewhere, but the queen was forgiving, it would turn out, as he lived to have further adventures, including exploring the coast of Maine and Penobscot Bay, before working his way down to Cape Cod.

Perhaps to Waymouth (and Elizabeth) the letter itself wasn't a priceless parchment. It was, if anything, a burning reminder of his failure…or, to be generous, a top secret government document that it was prudent to destroy. Why Lancashire? It's almost 300 miles from Waymouth's hometown of Cockington, Devon…but close enough to Liverpool to imagine he landed there and disposed of the letter along with any other of his expedition's detritus.

* * *

The second letter that could have altered the course of the Opium Wars was written more than 200 years later, in 1839, by an upstand-ing, albeit inflexible, bureaucrat named Lin Zexu. The Daoguang

Emperor had put Lin in charge of dealing with China's opium epidemic, and he was trying to appeal to the better ethical nature of the new queen, Victoria, as earnestly as Elizabeth had tried to appeal to the Wanli Emperor's better commercial nature two centuries before. However, whereas Elizabeth had looked for a way to make an alliance, the upright and formal Commissioner Lin, having come to the end of his moral rope, was threatening to end the fragile partnership the two countries had, by then, developed.

Lin seemed aware of the fact that communicating with a foreign ruler was way above his pay grade. Besides, he seemed more concerned with ingratiating himself with his own ruler than with Victoria. Clearly, the Chinese superiority complex at the time was a force as powerful as British noblesse oblige. "If there is profit," he explains, then his emperor "shares it with the peoples of the world; if there is harm, then he removes it on behalf of the world. This is because he takes the mind of heaven and earth as his mind."

He goes on to describe the British in terms he knew all too well were far from the truth: "By a tradition handed down from generation to generation [the English] have always been noted for their politeness and submissiveness." In another particularly grandiose passage he suggests that "we are delighted with the way in which the honorable rulers of your country deeply understand the grand principles and are grateful for the Celestial Grace." Then, he implies that the Chinese trade is the sole source of Britain's famous wealth, and goes on to make a barely veiled threat. He describes how some British subjects are seriously misbehaving, smuggling opium "to seduce the Chinese people and so cause the spread of poison." This, he claims, has put his boss in a "towering rage" and he was threatening to execute anyone, foreign or native, who sold or smoked opium in China.

Fortunately, he says reassuringly, Charles Elliot, the British government's chief superintendent of trade, had come to his senses long enough to agree to destroy more than 20,000 chests of opium, so he is confident the emperor would agree, in turn, to "magnanimously excuse them from punishment." But, Lin warns, the emperor will not be so forgiving the next time.

In closing, Lin becomes positively histrionic:

> "May you, O king,[2] check your wicked and sift your wicked people before they come to China, in order to guarantee the peace of your nation, to show further the sincerity of your politeness and submissiveness and to let the two countries enjoy together the blessing of peace....After receiving this dispatch will you immediately give us a prompt reply regarding the details and your circumstances of your cutting off the opium traffic. Be sure not to put this off."[3]

Some people say Lin sent the letter but Victoria's advisers kept it from her. Others believe it's equally likely he never even intended to send it. He simply had a few copies posted in the Canton Colony in order to threaten the merchants and impress the emperor. Regardless, he didn't wait for a reply.

---

2   He actually thought he was writing to a king, a mistake that is perhaps understandable despite the growing trade network between them, since Victoria had only taken the throne two years before. The idea that the English were ruled by a woman might have struck Lin as incredibly strange or even alarming, since his country's only reigning empress in history was the T'ang dynasty's Wu Zetian in the seventh century AD, who was said to have her own daughter killed so she could frame a rival.

3   Zexu, "Https://Cyber.Harvard.Edu/ChinaDragon/Lin_xexu.Html."

# Chapter 18

## Five Roads to War

While much has been written about the Opium Wars, there remains confusion about what happened and why, especially in the West where cultural biases lead people to believe that any war between China and the West that involved opium must involve the West trying to *stop* Chinese opium trade from dealing opium. Actually, if anything, it was exactly the opposite. The best way to understand the war and how it played out is to look at five intertwining narratives that describe the situation at the time.

The first describes the growth of the legal trade, which at times, depending on the current laws and regulations, might even include opium. Historians have spent years wading through millions of yellowing paper records, buried in government and company vaults, that document this trade—including the amount and price of every product, along with the tolls, tariffs, and duties to get it up the Pearl River.[1] Calculating those dealings alone could test the patience of the most stalwart abacus. And all the correspondence involved—in

---

1   A ship could include dozens of products from raw materials like lead, tin, and steel, to pearls, elephant tusks, and furs of beaver, otter, and African tiger. Sometimes they threw in a few "fish stomachs" (presumably they mean fish eggs. It's a little hard to read the old faded Dutch bill of lading). See Van Dyke, *The Canton Trade*, Plate 24.

which several letters could be exchanged simply trying to clear up a misunderstanding about the proper mode of address—do more to obfuscate than clarify the details. Regardless, it's clear that the amount of money and product changing hands in Canton grew steadily, and at times exponentially.

The second narrative describes how the *actual* trade worked on a daily basis. While a good deal was legal, as described above, the rest involved a tangled web of enticements, kickbacks, extortion, outright bribes, and the more blatant "connivances" that made it possible for higher officials to boost their salaries without much risk, since any time they were ordered to crack down on the illicit trade, there were always low-level smugglers to arrest or lower-level functionaries to discipline.

The third narrative thread follows the ever-changing patterns of outright smuggling (as opposed to official graft and corruption). At times, not only opium, but gold, iron, copper, salt, saltpeter, and even furs were heavily regulated or forbidden, and each ban helped a black market develop for delivering these goods illegally to connections on the mainland. Since opium was especially light and easy to transport, bags of it could be thrown into smaller boats while the rest of the shipment sailed merrily and legally along through the estuary's tollhouses to be weighed.

The fourth narrative describes the activities of pirates, who operated with remarkable impunity all over the South China Sea. While their delivery methods were virtually identical to that of the smugglers (whom they occasionally worked with), their techniques for obtaining product were even less savory.

Finally, way up north, in the capital of Beijing, there was a fifth narrative, the one that was *least* based in reality: the story that the emperor was telling himself about what was actually going on down there in Canton, primarily based on misinformation from advisers. Even when he did get accurate information about the opium trade, his conflicting priorities made it hard to decide what exactly to do about it or how seriously he should insist on enforcing his regulations. He knew that selling opium illegally made it possible for the British to afford his nation's tea; the fact that tea, unlike opium, loses its quality over time, meant he needed to be flexible in terms of the timing of any crackdowns. In addition, he wouldn't want to do anything that might seriously affect relations with foreign merchants because he earned a significant amount of money from the "emperor's present," which represented about 20 percent of every shipment and went directly into the royal treasury.

All these narratives were populated by a dramatis personae of unusual suspects, few of whom obeyed every rule or paid every tax. There were venal merchants and middlemen on land and sea, crooked ship captains, dope-smoking sailors, corrupt (as well as often dope-smoking) local officials, petty dealers, and the countless abused workers of both sexes and all ages who not only worked at the docks and in the warehouses of Canton but also back in India, cultivating, processing, and harvesting poppies under the watchful eye of East India Company officials.

The narratives would collide in unpredictable ways that changed year by year, season by season, emperor by emperor. One writer described the entire drama as something that "can best be compared to the parlour game called 'Cheating' in which everyone sets out to cheat without either scruple or blame in a language that no one really understood."[2] Even Shakespeare would have been pushed to his literary

---

2   Scott, *The White Poppy*, 18.

limits to combine the many narratives propelling the British and Chinese to war into one coherent plot—one that has many parallels to how hundreds of thousands of doses of illegal or unprescribed drugs end up on the streets of a nation that has defoliated foreign crops, jailed drug kingpins, invested billions in border security, and proved willing to establish mandatory minimum sentences for any citizen caught aiding or abetting the greedy foreigners and pharmaceutical companies trying to "push" those drugs on its people.

# Chapter 19

# The First Drug War

In the heart of New York City's Chinatown near the southern tip of Manhattan, at the intersection of Oliver Street, East Broadway, the Bowery, and Park Road,[1] stands a bronze statue of an imperious man, head turned right and gazing skyward, hands clasped behind his back, wearing a traditional Chinese robe and beads. The base is red granite from Xiamen in mainland China and the caption is simple: "Lin Ze Xu. Pioneer in the War Against Drugs." This is, indeed, Commissioner Lin who wrote the letter to Queen Victoria that was probably never sent. The letter that tried to convince her to end the British East India Company's opium traffic with China, so the two countries could "enjoy together the blessing of peace."

The fact that there's a statue in America of a Chinese bureaucrat who lived two centuries prior and was the pioneer in a war that most Americans think began in the late 1960s shows just how significant a role the Opium Wars play in Chinese cultural history.

Behind Lin, a block or so away, Confucius stands on a marble base from Taiwan. He faces the other direction, head tilted down instead of up; hands clasped in front of him, not in back. His caption

---

1  On some maps it's referred to as Kimlau Square and on others as Chatham Square.

is more detailed and is followed by a quotation from his essay *The Great Harmony* or the *Ta Tung,* in which he describes an ideal society, in which "selfish schemings are repressed, and robbers, thieves and other lawless men no longer exist." This was clearly not the environment Commissioner Lin found himself in when he arrived in Canton. In spite of increasingly strict laws that included closing opium dens and summarily executing some dealers, the "selfish schemings" of dealers, thieves, and corrupted officers continued virtually unchecked. Opium use wasn't simply an issue of morality now for the Chinese government: the drug was also having a serious effect on the nation's productivity, since millions of addicts from all walks of life were spending some or all of their days eating, drinking, and smoking. (Some estimated that 20 percent of all government officials smoked opium, and a majority of the merchant class did.) On the other hand, Great Britain needed to *increase* its opium trade because independent merchants had recently broken up the monopoly of the East India Company. This ushered in a new era of competition, which led to lower prices. Since income from opium sales played a major role in Britain's ability to buy tea, the lowered prices meant they needed to cultivate more addicts in China.

The interplay between supply, demand, price, and tariffs was extremely fluid. While limited opium imports were still legal, they were heavily taxed. Some of the emperor's advisers thought the best solution, therefore, was to turn the screws harder on regulation so as to squeeze every possible tax dollar out of each shipment in order to fill the royal coffers, while weaning the population off the drug. However, the statuesque Commissioner Lin, the leader of the vigilante anti-opium faction, argued that, at this rate, there soon wouldn't be any peasants left to power the economy; students would fall asleep on their desks, and soldiers wouldn't just abandon their posts, they wouldn't report for duty in the first place. For Lin, the time had come to destroy the opium coming onto Chinese shores, execute or banish all Chinese dealers, kick out the worst of the English merchants in Canton, and, from then on, refuse to allow any boat carrying opium from entering the Pearl River estuary under any circumstances.

Lin's opinion carried a lot of weight with the Daoguang Emperor. He had submitted a detailed report on the opium trade that covered everything from the players to the rituals and paraphernalia involved to the best alternative drugs and herbs for those going through withdrawal.[2] Recently, he'd been governor-general of two provinces where he had allegedly ended opium use while making significant achievements in flood control, social services, and tax collection.[3] He knew how to get things done. Lin seemed like the perfect choice to send into this world of trickery and deceit. He was intelligent, of unimpeachable character, unwavering in his resolve, and experienced when it came to dealing with groups who had competing interests.

Lin knew he had to make it perfectly clear that if companies were interested in continuing to do *any* business at all in China, they had to stop selling opium once and for all. On March 10, 1839, he arrived in Canton and after appropriate pomp and circumstance immediately put his plans into action.

First, he ordered the street around the Canton Colony (foreigners' zone) to be barricaded, the river to be blockaded, and had more than 70,000 opium pipes confiscated. Then he used his expertise in flood control to set an army of laborers to work digging and lining trenches seven feet deep, 25 feet wide, and 150 feet long. Seeing this, the British merchants urged trade superintendent Captain Elliot,

---

2    Waley, *The Opium War Through Chinese Eyes*, 24.
3    Hedman, "Commissioner Lin Zexu and the Opium War."

who represented their interests in China, to defend the dignity—and bottom line—of Her Majesty and Her Subjects. Elliot sailed up the estuary from the South China Sea and began a series of tense negotiations with Lin.

Captain Elliot was no admirer of the opium trade, and he had already warned his British countrymen that they needed to end their smuggling practices before China lashed out. Ultimately, he decided the best course was to agree to Lin's demands to hand over any opium on British ships currently in Canton harbor and to instruct any ship at the mouth of the Pearl River estuary in Macau to quickly unload any *legitimate* cargo, and leave the area.

The British merchants would have been loath to hand over their opium any time of year, but this was the season's first crop, which was the time when they most needed revenues to help finance the cultivation of subsequent crops back in India. It was only Elliot's optimistic assurances that, somehow or other, their losses would be covered—as well as his warning that even their *legal* product might be seized—that convinced the English merchants to finally give in.

Elliot's move was calculated, controversial, and complicated, perhaps even more than he realized. In the short term, he was saving significant *legal* trade from being confiscated, but, in the process, he was agreeing to surrender British goods to a foreign country that would proceed to destroy them. At the same time, the situation was eerily reminiscent of what had happened in 1773, the last time Britain hadn't compromised on trade issues. In that case, American colonists had confiscated British tea and unceremoniously dumped it in Boston Harbor. So, while he knew that allowing China to destroy British goods set a terrible precedent for the future of the China trade, he feared that a confrontation could be even more destructive in the long term.

Still, he stalled for a while. The ultimate diplomat and pragmatist who was representing the new, still teenage, Queen Victoria—a job he didn't take lightly—Elliot was especially reluctant to misstep. He first tried explaining to Lin that England wasn't even to blame for the problem, using the familiar excuse that it was India that was really pushing the opium. (Although since India was largely a British colony by then, this argument wasn't very convincing.) Then he told Lin that he needed to wait for instructions from back home and asked if, until then, ships currently in Canton Harbor could remain safely anchored there. Lin and the Hoppo considered this to be a ridiculous delaying tactic, and to make their point, they subjected Elliot to a rash of self-righteous Confucian posturing about the correct behavior between people and countries.

Through it all, Elliot was involved in constant negotiations with the foreign merchants of Canton, who became increasingly impatient about the money they were losing every day as long as trade was at a standstill. At times, it seemed that no one knew exactly what was going on, what they were supposed to do, or whose side they were on. There was a sense that the smallest misunderstanding could easily spark a confrontation. It almost happened when a British ship near Macau fired a shot toward a Chinese war junk during some hijinks as they celebrated Queen Victoria's twentieth birthday (May 24, 1839). Fortunately, that junk moved away from the British ship and no one was injured.

By early June, Lin's trenches were complete. He had already confiscated 3 million pounds of opium, and he was ready to begin destroying it. First, he checked with Beijing as a formality to make sure the emperor didn't want to destroy it himself in the capital. To his surprise, he agreed at first, until more perceptive heads prevailed and pointed out to the emperor how much pilfering would likely occur during the long journey to Beijing.

With permission secured, Lin ordered the workers to fill the first trench with opium and then throw in salt and limestone, transforming what a famous physician would one day call "God's own medicine" into what Lin referred to as "foreign mud." As the story goes, Lin raised his hands to the heavens, asked forgiveness from the spirit of the sea, and urged all sea creatures to retreat to deeper water, lest they become hopelessly addicted. The sluice gates were lifted until, slowly, the muck began to slide through a screen into a creek and disperse into the bay. Undoubtedly, there were many users in the audience who would have preferred that he set it on fire instead so they'd at least be able to enjoy some secondhand smoke. Some historians argue that's exactly what he did do.

Since he had caught wind of rumors that he wasn't actually going to destroy all the opium, Lin agreed to let a blameless merchant— one of the few—named W. W. King and his wife see for themselves how he was pulling off this act of mass destruction. Accompanied by a guide named Loo, they and several others, including a missionary named Elijah Bridgman, arrived in Canton Harbor to witness a great celebration with crowds of spectators on boats and in nearby houses who looked expectantly at a 400- to 500-foot square enclosure, protected by a fence made of bamboo stakes pounded deep in the ground. Bridgman later described the scene in his magazine the *Chinese Repository:*

> One trench was empty, another one was being filled by coolies who carried the opium in baskets, put the balls on planks laid across the trench, and stamping on their heels to break it into pieces, before kicking it into the water....Other coolies were employed in the trenches, with hoes and broad spatulas, busily engaged in beating and turning up the opium from the

bottom of the vat. Other coolies were employed in bringing salt and lime[stone], and spreading them profusely over the whole surface of the trench. The third was about half-filled, standing like a distiller's vat, not in a state of active fermentation, but of slow decomposition, and was nearly ready to be drawn off.[4]

According to Bridgman, King was particularly impressed with how well the entire process was choreographed to prevent any of the workers from absconding with so much as a thimbleful of opium. Satisfied, King and Bridgman went off to meet with Commissioner Lin, after Loo assured them the ladies would be served afternoon tea and sweetmeats.

Arriving at Lin's office, they found him waiting patiently beside the commander in chief of the Chinese navy, the current Hoppo, and the commissioner of justice. Over the next couple of hours, King carefully pointed out that a lot of money had just gone down the drain, so it was important to reestablish trade, especially his own, as quickly as possible. Lin convinced him there was nothing to worry about. His twin goals, he said, were aligned with theirs: to stop the opium trade and relaunch legitimate business as soon as possible. On the subject of any reparations, Lin remained inscrutably mum.

They discussed other issues, including the best way for Lin to get messages to that young British queen, new schemes for taxing imports, and whether British criminals could be tried in Chinese courts. Overall, King seems to have been pleased with the whole encounter and even somewhat relieved that there was nothing "barbarous or savage" about Lin.

---

4    Bridgman, *The Chinese Repository VIII*. Also see Waley, *The Opium War Through Chinese Eyes*, 49 ff.

The laborers continued refilling the trenches with opium for ten days, in what was presumably the largest destruction of a controlled substance in history.

The British weren't the only ones who were financially concerned. The Hoppo himself, and the rich and powerful hong brokers, couldn't afford to have boats with legal product floating around the sea while Lin determined how he could confirm they didn't carry opium or whether he would keep them out of Canton altogether. Lin's ban was also a serious problem for the Chinese officials who supplemented their modest salaries with bribes, not to mention the people obliquely involved in the opium economy, such as the artisans who crafted the elegant pipes and accessories that made smoking as pleasurable as possible or the tens of thousands of ordinary citizens whose lives depended on the China trade, whether legal or not.

Major changes in any government's approach to drug enforcement can have unexpected economic consequences. In this case, the people who most benefited from Lin's destruction would prove to be smugglers, who eagerly watched their margins go through the roof as the price of opium skyrocketed, while looking warily over their shoulders, knowing that the cost for being caught was going up as fast as the price for the product itself.

Even after ten days of destroying the drug, Lin wasn't satisfied. He ordered innkeepers and landlords to file reports every five days listing the names of everyone who was moving in and out of town. He established security groups for officials, soldiers, and clerks—partly to make sure they still had some work to keep them busy. He continued to execute or imprison dealers and forcefully hospitalize users for treatment, based on information provided by informants—some of whom used the moral panic as an opportunity to rid themselves of rivals or annoying relatives. He even arrested certain *native* merchants whom he considered traitors to the country.[5]

Elliot wasn't done yet either. While some impugned his strategy at the time, and some historians still do, he was taking the long view. As significant a loss as the opium represented, it gave him a leg to stand on, albeit a shaky one, as he prepared to fight back. Instead of trying to defend the British right to trade a forbidden narcotic, he framed Lin's public destruction as a seizure by one sovereign government of goods belonging to another foreign government. As soon as word spread in London that the British flag had been insulted, Lord Palmerston, the secretary of state of foreign affairs, agreed to call in the navy.

In the standoff between Lin and Elliot, the British had the upper hand in a couple of important ways. One was that, as his letter to Queen Victoria demonstrated, Lin didn't really understand how to negotiate with the British. He had read up on them in less than reliable sources, one of which suggested that they made their opium by mixing poppy juice with human flesh, although Lin argued it was actually mixed with the corpses of crows. The confusion came about, he explained, because "foreigners expose their dead and let the crows peck away the flesh. That is why the crows shown in pictures in foreign books are of such enormous size, sometimes being several feet high. Consequently, they could certainly obtain sufficient flesh from crows, without having recourse to human flesh."[6] (Lin was referring to the ancient Middle Eastern Zoroastrian custom of disposing of the dead by putting their bodies on the roof to, indeed, be eaten by

---

5   Descriptions of Lin's enforcement tactics range from these kinds of indiscriminate arrests and/or executions based on questionable reports (see Hedman, "Commissioner Lin Zexu and the Opium War") to a relatively enlightened approach in which, basically, dealers were punished and users were offered treatment.

6   Waley, *The Opium War Through Chinese Eyes*, 67.

crows—which made sense in a desert land where burial was uncertain and firewood was in short supply.)

The second advantage the British had—and one Lin wouldn't have yet been able to fully appreciate—was that the British navy had made significant technological advances in the preceding decades. Their firepower far surpassed that of the many ships with which the Chinese guarded their harbor. Even thousands of miles from home, Elliot's ships could dominate the Chinese.

The next several months progressed in a swirling mass of threats and counter-threats, accusations and counter-accusations, affronts and counter-affronts. After Elliot refused to hand over two British sailors who, drunk on rice wine, were accused of killing a man in the town of Kowloon, Lin told his people to stop selling food to the British on ships waiting at the mouth of the estuary.

For Elliot, this was the final straw. It was one thing to have the Chinese destroy his country's product, but to try to starve out its people was an act of aggression. He issued an ultimatum that Lin immediately lift the blockade on food supplies. When Lin refused, the British opened fire. The Chinese shot back, both from land and sea. Still, there was no serious damage to either side. Lin allowed the British to buy provisions again.

At that point, he offered to let any boat willing to sign a bond that its cargo was opium-free to go upriver and unload at Canton again. As far as Elliot was concerned, putting British merchants in a situation that required them to sign an official agreement with a foreign government was an unacceptable attempt by China to usurp the British government's ultimate authority over their own merchant ships.

Perhaps it is fitting, given the complex internal politics that set Britain and China on course for the conflict, that the first serious battle in the Opium Wars started in November 1839—less than a year after Lin's arrival—when the British ship *Volag* fired a warning shot across the bow of another *British* ship, the *Royal Saxon*, because its captain had signed Lin's bond—against Elliot's instructions—and was proceeding up the Pearl River to unload its opium-free cargo at the port of Canton.

In response to those British-on-British shots, some Chinese war junks, five of which happened to have red flags—typically considered a declaration of war in the West (although not in China) went out to protect the *Royal Saxon* as it tried to outrun its *own* country's virtual blockade of its *own* ships from another country's harbor. In response, the *Volag*, joined by another British warship, the *Hyacinth*, opened fire, blowing up two of the Chinese junks, sinking three others, and wreaking havoc on the rest. All told, fifteen Chinese were killed and one British sailor wounded that day.

In the wake of the violence, the Chinese retreated to nurse their wounds. Lin tried to put the best face he could on the humiliating defeat and convince the emperor that their confrontational strategy was beginning to work.

By the summer of 1840, the British had slowly but surely begun to take control of every nook and cranny of the Pearl River estuary, and its boats were delivering cargo—whether legal or not—to any merchants along the coast willing to risk Lin's ire. By the summer of 1841, they had captured a large swath of southern coastal China, and the emperor, seeking someone to blame, sent Lin into exile for four years. There, Lin confided to friends that he knew defeat was inevitable after he failed to convince the emperor that they needed to significantly ramp up their forces once he had fully realized the West's overwhelming naval superiority.

By the summer of 1842, the Chinese had agreed to the lopsided Treaty of Nanking, which awarded the island of Hong Kong to the

British and gave them free access to Canton and Shanghai along with five new ports farther north. It also ended the hong merchants' stranglehold on trade, established standard, predictable duties, and included $21 million to cover the costs of, among other things, the reimbursement that Elliot had unsuccessfully demanded for the confiscated opium that Lin had destroyed three years before.

Finally, as if the British government couldn't help underline its newly acknowledged dominance, the treaty required that official correspondence between China and England reflected the countries' equality instead of any Chinese insistence that the British kowtow when making requests.

In October 1843, the British insisted on additional provisions: first, to ensure that when their citizens were accused of crimes in China, they would be subject to British, not Chinese, law; and, even more important, to secure Britain's most-favored-nation status, which meant, essentially, that the Chinese couldn't give any country better prices or lower tariffs than England's.

The emperor was outraged at these deals, which were signed by Lin's pragmatic successor, Qishan. The exiled Lin was soon allowed to return while Qishan got his turn to lose his property, his house, and even his concubines before being sent off into indeterminate exile.

When the dust settled, the smoke cleared, and the negotiations ended, neither the war nor the treaties had altered the course of the opium trade. In the end, the treaties didn't so much as mention opium, and they didn't need to. They granted British companies license to trade any goods they wanted at the ports. As far as they were concerned, the Chinese government had the right to outlaw opium, punish native users, and even execute native dealers if they so chose. In other words, as far as the British were concerned, they could police the citizens in their nation however they wanted. The merchants knew that opium addiction had taken root, and no matter what Chinese leaders did to punish their people, there would always be more buyers on the streets and more traders ready to buy shipments wholesale, and make a profit by selling the product to small dealers or individual users.

By the time the ink was dry on the Treaty of Nanking, opium imports were up 20 percent from before the war. Within two decades, they would double.[7]

Almost two centuries after the Opium Wars ended, historians continue to fight over their causes and implications for today's global conflicts. Some blame the Chinese, saying disagreements among its leaders (some of whom used the drug itself) about the best way to keep foreigners out while profiting from the trade made conflict inevitable.[8]

However, the noted British historian Nicholas J. Saunders called the Opium Wars among the most immoral episodes in his country's history, concluding, "These conflicts saw the British Empire officially trafficking opium, and using military might to force narcotic addiction on the people of China. During these years, Britain created the largest, most successful and most lucrative drug cartel the world had ever seen."[9]

Regardless, repercussions from the Opium Wars continue to resound today. As Julia Lovell, another British expert in Chinese history put it: "From the age of opium-traders to the Internet...China and the West have been infuriating and misunderstanding each other....Ten years into the twenty-first century, the nineteenth is still with us."[10]

---

7   Extrapolated from Pietschmann, Tullis, and Leggett, "A Century of Inter-national Drug Control."
8   O'Connor, "We're Still Fighting the Opium Wars."
9   Saunders, *The Poppy*, Ch. 3.
10  Lovell, *Opium War*, 361.

# PART V

## The Agony and the Ecstasy

# Chapter 20

# America Enters the Opium Business

Although the United States was a much smaller player than Great Britain in the opium trade at first, American companies would eventually use equally shameless rationalizations and brazen self-interest to invest in it and, later, collective amnesia to avoid acknowledging the damage it caused. That duplicity, combined with racial prejudice, has led to the assumption that continues to this day, that America's opium problems in the nineteenth century were caused by China when, if anything, it was the opposite.

A few American ships had traded in the East prior to the Revolutionary War, but always under the auspices of Great Britain. As the Boston Tea Party famously proved, by the 1770s, tea had become an important commodity in America. Its 4 million citizens were now drinking more than 1 billion cups annually. They needed to replace that supply as quickly as possible.[1]

---

1   It has been suggested, scurrilously or not, that one reason John Hancock and his fellow rabble rousers dumped tea in Boston Harbor was that they were tea smugglers. The British Parliament had passed the Tea Act in 1773 to save the failing East India Company. Rather than raising tea taxes, as usually assumed, it actually *lowered* them so the East India Company could be competitive again. This was really bad for business for tea smugglers like Hancock. So, they dumped the tea to reduce supply and raise prices for everyone. See Chapman, "Taking Business to the Tiger's Gate." The "1 billion cups" figure is Dolin, "How the China Trade Helped Make America."

On February 22, 1783, a 150-foot-long three-masted ship called the *Empress of Philadelphia* sailed from New York harbor, marking America's independent entry into the Far East trade. It could hold only 360 tons, less than half the capacity of most East India Company vessels, but was a more versatile, practical, and cost-effective size for the early days of the America-China trade.

The *Empress* was loaded with cordage, cloth, silver, beaver furs, and nearly thirty tons of wild ginseng that had been harvested from the woods of southern Appalachia (primarily by the Cherokees, who knew where it was hiding and the delicate techniques for harvesting it, knowledge they jealously guarded).[2] While the British could dig into their coffers of silver to make up any difference in their balance of trade with China, America wouldn't have any silver until they started mining the Comstock Lode in Nevada in 1858. Until then, ginseng played a critical role in the China trade, since the Chinese used it to treat many ailments associated with weak constitutions, including impotence, and were willing to pay top dollar for it.[3]

Fundraising for the *Empress*'s voyage was spearheaded by Robert Morris, a largely forgotten signer of the Declaration of Independence who, like everyone else involved in the shipping industry before the Revolutionary War, was scrambling to establish new trading opportunities. Morris was one of the merchants who lent the government money for warships. Refitting them for peacetime work

---

2   The natives in Canada had made the "mistake" of showing French fur traders how to find and dig it, and Old World settlers down in the States also caught on pretty quick. Daniel Boone, for example, sent a whole barge full of it to Philadelphia in 1787. That batch got ruined when the boat got swamped, so he just sent his diggers (a.k.a., "sangers") back for more.

3   In animal tests ginseng has proven that it can treat erectile dysfunction without side effects. Lee and Rhee, "Effects of Ginseng on Stress-Related Depression."

was a win-win: it would let the merchants start making money, while enabling the government to begin collecting import duties again. While some accused Morris of being a war profiteer, others considered him the financier of the American Revolution because he begged, borrowed, and gave so much money outright to the cause—including funding the purchase of 80 percent of all bullets the Americans fired during the war. He also worked closely with Alexander Hamilton to develop a stable treasury for the new country and urged George Washington to appoint Hamilton to be its first secretary.[4]

Morris and his partners remained hesitant about risking too much on trade with faraway China until they consulted with a Dartmouth dropout named John Ledyard.[5] In the course of his restless travels with the British explorer Captain Cook, Ledyard had seen the exorbitant price the Chinese were paying for opium.[6] Even with Ledyard's advice, however, the letter of introduction that the American merchants gave the captain of the *Empress* proves they had little idea as to the proper way to go about trade in China (or even whom to go about it with). The letter begins:

---

4 Morris was both a patriot and a profiteer during the war. Several times he paid the troops from his own funds and provided his own ships to support the war effort. At the same time, he was able to keep some of the goods that the colonists seized from British ships and made a profit selling them in France and Spain. By the end of the war, while he had lost one of the largest private navies in the world, he never asked for reimbursement from the new government. In the end, he figured that he came out about even. See: Invaluable (Auction House), "Morris, Robert. Autograph Letter Signed, 1 May 1776."

5 After dropping out of (or being expelled from) Dartmouth, Ledyard hollowed out a log (undoubtedly with the help of some friendly native Abenaki) and floated down to his parents' house in Hartford. Their reaction to his floating up to their door is, unfortunately, not recorded.

6 See Hamilton, "Five Fascinating Facts about Sea Otters."

Most serene, serene, most puissant, puissant, high, illustrious, noble, honorable, venerable, wise, and prudent Emperors, Kings, Republics, Princes, Dukes, Earls, Barons, Lords, Burgo-Masters, Counsellors, as also Judges, Officers, Justiciaries and Regents of all the good cities and places, whether ecclesiastical or secular, who shall see these patents or hear them read.[7]

The letter goes on to ask in the most polite terms to be taken seriously. Interestingly, one of the Americans' errors was *not* coming bearings gifts, which bewildered the Chinese if it did not insult them outright.

The *Empress* took off from New York Harbor with a forty-two-man crew and a thirteen-gun salute as if it represented the patriotic hopes of the thirteen new states and not just the financial hopes of Morris and his associates. It was a bold experiment for the Americans. If not for a recently published set of maps called *A New Directory for the East China Sea,* the captain would have had a very difficult time navigating all the way to Canton. Fortunately, he also had help from the French, America's allies in the war against Great Britain. Two French ships rendezvoused with the *Empress* in Indonesia's strategic Sunda Strait, protecting them from pirates as they guided them all the way to the Pearl River. The Americans arrived on August 23, roughly six months and 12,000 miles after they left.

Unfortunately, Morris and his partners had failed to account for the fact that their arrival with the largest shipment of foreign ginseng ever to arrive at a Chinese port would drive the price down, which dashed their hopes of making a huge profit. Still, Morris & Company did

---

7   Dolin, *When America First Met China,* 20.

reasonably well selling the furs, and returned to America with a huge amount of tea, piles of fine cotton trousers, and some equally high-quality porcelain. When all was said, done, and tallied, the voyage earned a 25 percent profit for its investors, and America's merchants had learned a great deal about the logistics of long-range sea trade and how to negotiate profitably with the Chinese.[8]

---

8    The rest of Morris's career in commerce was, sadly and unfairly, not so successful. His next attempt to help get the country and himself back on their financial feet was to make a deal with France to sell them a serious amount of tobacco every year, only to have Thomas Jefferson (who undoubtedly had his own tobacco interests to consider) get in the middle and collapse the market. Morris was then sued by a Virginian tobacco grower and eventually forced into bankruptcy by some ill-advised land speculating he did to try to pay the guy off. He ended up spending three and a half years in debtor's prison in Philadelphia and retired from public life. Robert Morris University in downtown Pittsburgh—which began as Robert Morris School of Business—is named after him in recognition of his reputation for having financed the Revolution.

# Chapter 21

## Generosity and Greed

Two centuries before the Sackler family started endowing nonprofit institutions by using profits from Purdue Pharma's aggressive and deceptive marketing of the addictive prescription painkiller OxyContin, distinguished families in Boston were using opium profits to do much the same thing.

Among the famous institutions that benefited from the opium trade is Boston's Museum of Fine Arts, one of the largest museums in the United States. Buried in its vaults is a nineteenth-century oil painting by an unknown artist entitled *Hongs at Canton, China*. Its romantic, sanitized version of Canton Harbor is a perfect example of the nineteenth-century self-mythologizing that obscures America's complicity in the Chinese opium trade. It shows a row of buildings that look like Washington, DC, embassies sitting above docks, with small sections of browning lawn sloping from them down to the water. To the left and perpendicular, rows of identical warehouses move up the canvas. They look like tobacco barns, although, in this case, you can almost smell the opium. The water itself is an emerald green reminiscent of the Caribbean or deep limestone pools. A few dozen oared boats crowd around the docks, while out in the water there are several one- and two-masted ships, two with sails furled, as well as a few other junks and a houseboat. In defiance of the conventions of perspective, the boats in the *foreground* are *smaller* than the same-size ones in back, insistently drawing the viewer's eye to the six flags beyond the harbor itself, which fly on high poles next to the embassy-like buildings. From the left, they are Denmark, Spain, United States, Sweden, Great Britain, and the Netherlands. Five of the flags represent companies that, like the British East India Company, were monopolies, their rights to trade owned, or at least granted, by their governments.

The United States flag, however, represents a half-dozen independent American trading companies that share the headquarters. The fact that America's embassy is in the middle may say more about the artist's national bias than the actual arrangement, and the fact that there are only a few sleepy ships in the harbor shows significant artistic license since, while the foreign headquarters were elegant, thousands of boats crowded the docks and banks of the river day after day, and the warehouses were stacked to the roofs with loads of fetid furs, wool, silver, and opium, surrounded by sweaty bean counters tallying endless figures while wondering if they were ever going to able to return to civilization.

The buildings and warehouses look fairly new. They were, in fact, built by hong merchants who rented them out to their foreign trading partners. While the American companies were competitors with each other, they were for the most part friendly ones. In the best of times, they cooperated, sharing storage facilities, lending and borrowing money, and entertaining themselves as best they could. But during the week they could be ruthless competitors.

Regardless, almost all would make a *lot* of money.

Some of that money would help fund the Boston Museum of Fine Arts, home of the deceptively romantic painting *Hongs at Canton, China*.

Another venerable American institution that benefited from Boston's involvement in the opium trade is Massachusetts General Hospital, the third-oldest general hospital in the country. Just down Huntington Avenue from the Museum of Fine Arts, the hospital has to deal with the opioid epidemic on a daily basis. Its Center for Addiction Medicine provides treatment, research, and education in a multidisciplinary outpatient setting. While many emergency rooms are only equipped to *revive* addicts who have overdosed, at Massachusetts General's ER *anyone* suffering from an addiction—whether in crisis or not—can receive immediate medication-assisted treatment for their illness (e.g., Suboxone) just as they would if they'd arrived with a fractured bone, shortness of breath, or inexplicable sharp pain.

While the hospital was incorporated in 1811, construction didn't begin until 1818. Donations ranged from 25 cents to $20,000, with a 273-pound sow of indeterminate value in between. Several of the $20,000 gifts came from respected Boston Brahmins who were busily making that money and more selling opium in China.

In addition, McClean Hospital—one of the earliest and foremost psychiatric centers in the country—owes its 300-acre facility in Belmont, Massachusetts, in part to drug money. In the 1950s and '60s, several famous writers with drug problems spent time there, including Sylvia Plath, author of the seminal *The Bell Jar,* which is a thinly veiled description of her time as a patient at McLean. The poet Anne Sexton was also a patient at McClean and would, like Plath, commit suicide. Her poem "The Addict" is a tragically revealing look at the experience of mental illness, drug addiction, and suicide ideation.[1]

While at McLean, Sexton taught her fellow inpatients a Ken Kesey–worthy poetry seminar. Among the student-patients was a young man

---

1   Sexton, *The Compete Poems,* 165.

named Robert Perkins. His great-great-great-grandfather or uncle was the philanthropist Thomas Handasyd Perkins, one of McClean's founders and the most successful and unrepentant drug dealer in Boston high society during the early 1800s.[2]

\* \* \*

Thomas Handasyd Perkins had been chomping at the bit to get in on the opium action since he first heard about merchants in Philadelphia and Baltimore who had made fortunes selling the high-grade Turkish variety in China. Perkins was ideally positioned to do the same: he had family members in Smyrna on the Greek coast across from Turkey who could provide him with inside information on the supply side so he could accurately forecast price fluctuations. Plus, his nephew John Perkins Cushing was already in Canton handling their existing non-opium trade, which involved perfectly innocent loads of cheese, lard, and iron, which they traded for equally legal loads of tea and silk. With his experience and resources, it didn't take long for him and his extended family to emerge as the dominant American merchants.[3]

In addition to their global connections, the Perkins family was helped by a number of other factors. First, they had an appetite for risk and a sense for the ebb and flow of Chinese enforcement—in fact, they enjoyed watching their more risk-averse competitors get scared out of the business during the regular Chinese crackdowns. They also had the money, nerve, patience, and strategic insight to survive the

---

2   Thomas's partner in most business and philanthropic enterprises was his brother James who played a less prominent, albeit equally lucrative, role.

3   Some sources claim the first Perkins opium shipment didn't happen until 1816, which is kind of hard to believe considering how quickly they established their dominance. See "Narco-Philanthropy."

cycles of booms and busts in the market for Turkish opium—some of which they even caused intentionally. Even the War of 1812 didn't faze them. One time during that war, a Perkins ship captured two East India Company ships and stole all their gold and silver, proving that they were more than willing and able to blur the line between profiteering, piracy, and patriotism.

By the 1820s, they dominated the supply chain, being virtually the only American company left standing in the Turkish opium market. Even the British were too preoccupied with their production in India to challenge the Perkinses' power in Turkey.[4] They would soon realize, however, that someone with only a fraction of the market could manipulate prices in Canton as well as they once had.

Eventually, the Perkins Company had 200 vessels carrying opium, tea, silks, spices, porcelain, and other goods to and from ports in the Far East as well as back and forth to America. The company was always looking for ways to innovate. For shorter runs up and down the coast, they started using a new kind of boat called a clipper, which was narrower, smaller, and a lot faster than traditional boats—designed not only to move product more quickly, but also outrun Chinese officials in pursuit, as they bought and sold opium at ports along the coast where they were forbidden to trade.

Meanwhile, in the early decades of the nineteenth century, the entire Pearl River region grew into something that resembled a giant railroad switching yard.

---

4   Britain's other distraction was the fact that, even though they'd taken control of *Patna* opium, named after one of the major cities in northeast India, the states in central and northwest India that were still independent were increasing production of equally good and less expensive *Malwa* opium, which *anyone* could buy. Eventually, the British contented themselves with charging the free states a healthy transfer tax in exchange for access to their more convenient port of Bom.

Juggling all the products flowing into and out of that region was an incredibly complex task, the equal of the action on the most sophisticated modern-day trading floor, albeit at a slower pace. Yet despite the many people involved, remarkably few were in the direct employ of the Perkins family, and even fewer were involved in making day-to-day operational decisions. Somehow, Thomas Perkins found a way, in an era before even the telegraph was invented, to manage this astonishingly complicated system from thousands of miles away—setting limits on prices and volume for his buyers in Turkey, determining the best time for his agents in Canton to sell the latest shipment, and perhaps even conducting a form of industrial espionage to gather information about his competitors.[5]

During the first few decades of the 1800s, Perkins's business evolved and expanded thanks to marriages and strategic alignments, until he could boast the greatest resources, best commercial intelligence, most reliable ships, and most savvy financial strategists in Boston, London, Smyrna, and Canton.

The second-most influential member of the Perkins family was Thomas Perkins's nephew, John Perkins Cushing, who would become the most powerful merchant in Canton and, in the process, teach the tools of the trade to yet another generation of impressionable young relatives. Taken in by his uncle when his mom died, Cushing was sent to Canton in 1803, at age sixteen, to learn the trade from Ephraim Bumstead, the head of the Perkins house in Canton. When Bumstead died suddenly, Uncle Thomas could have sent someone experienced to take over, in order to give John time to continue his apprenticeship. But, since Canton was still a relatively small part of Perkins's overall

---

5    For a good overall description of the Perkins opium trade, see Chapman, "Taking Business to the Tiger's Gate."

trade back then, he and the family decided to see how their precocious nephew could do on his own.

Not only did Cushing keep his head above water, but he is said to have ended up with the most money of all—much of it made *after* having publicly disavowed all involvement with the opium trade, which he did in 1818 to stay on the best possible of terms with Howqua, the most famous, wealthy, and powerful hong agent in Canton. The arrangement was something of a charade, since even though Howqua frowned upon the opium trade, he always found ways to benefit from it. For example, when Cushing disavowed the opium trade, he announced that henceforth his Canton commission business would be handled by the J.P. Sturgis & Co., which, it may go without saying, was run by three partners who were also nephews of Thomas Perkins.

Neither Thomas Perkins nor John Perkins Cushing was ever truly out of the opium trade, as indicated by a letter that Thomas sent to his nephew in 1826. In it, the elder Perkins apologizes for the fact that 150,000 pounds of opium was on its way from Turkey to Canton (four or five times as much as usual). He claims to be thoroughly mortified that he was sending far more opium that season than Cushing had asked for but said that he felt it was necessary in order to maintain their monopoly in Smyrna.[6] It's hard to imagine Thomas Perkins's apparent "groveling" as anything other than an amused tweak at his nephew— if not a double-down-dare to see if he still had the chops to move that much product while pretending to be out of the business. One way or the other, he did. 1826 proved to be a good year for the company.[7]

---

6   Scholars have spent a lot of time trying to get the numbers right on the opium trade. This number, however, could have represented virtually the entire Turkish crop for that season—as much as four or five times what Cushing was used to receiving. See Downs, "American Merchants and Opium Trade."

7   Bebinger, "How Profits from Opium Shaped 19th-Century Boston."

\* \* \*

Since the members of the Perkins family never married below their station and had no station to marry above, over the next few generations many of the most famous names in Boston society would be swept into their world of wealth, generosity, and questionable morals. There's no need to parse the dizzying network of connections that was their family tree, loosely known as "the Boston Concern," except to recognize that they owed much of their financial success to the opium trade—to the dismay of some of their children and virtually all of their descendants.[8] There is, however, one player who deserves special mention—Warren Delano II—some of whose money would end up in the accounts of one of America's most famous presidents. Born in 1809, he began working in the trade in Boston as soon as he was old enough to make money. While not a blood relative of the Perkinses, he was a contemporary of Robert Bennet "Ben" Forbes, and eventually they became junior partners in a company called Russell, Sturgis & Company, which through a series of mergers, buyouts, and family agreements ultimately became Russell & Company, which by 1840 was still managing the opium trade on behalf of Perkins, and was the largest American firm in the China trade.

Having made $200,000 (twice what the family considered needed to prove a "competence") Delano retired and returned home at thirty-one to live in the manner to which he was born. When his fortune took a beating in the Panic of 1857, he went back to China to see if his luck—and skills—still held. They did. Boosted by the rising price of

---

8   A good number of Bryants, Paines, Higginsons, Wilcocks, Willings[es], and
    Latimers were also involved in the trade, so, if any of those names appear in
    your family tree, you too might be a beneficiary. Downs, "American Merchants
    and Opium Trade."

opium that resulted from the huge demand to treat wounded soldiers during and after the Civil War, Delano died a wealthy man again, but not before marrying Catherine Robbins Lyman, who begat Sara Ann Delano, who begat Franklin Delano Roosevelt.[9]

One reason that Americans such as Perkins, Cushing, and Delano were able to avoid prosecution, especially after the First Opium War— regardless of how questionable their business practices were—was that the only laws that American merchants broke were ones made in China. However, the Treaty of Nanking included provisions that any American merchant or associate accused of breaking a Chinese law would have to be prosecuted by Americans. And, since America didn't yet have any laws about trading the drug, they, essentially, couldn't be prosecuted at all. In fact, it would be 100 years before there were any sort of agreements on extradition for drug crimes committed in one country by a citizen of another. And they didn't work out so well either. The worst that could happen to merchants—and occasionally it did—was that they'd be asked to leave the country and their property would be confiscated.[10]

As far as any moral or ethical laws the upstanding, churchgoing Perkinses were aware (or cared) they were breaking, they undoubtedly justified them based on reports that, in China, opium was no better or worse than alcohol. (It was certainly better than the slave trade Perkins had profited from early in his career.) Besides, they could always fall back on the eternal law of questionable trade—someone's going to make money on it, why not them?

---

9   Ward, Geoffrey, *Before the Trumpet*, Ch. 2.
10  In the short term, sovereignty was far more important for drunken American sailors who accidentally killed Chinese citizens (which happened several times). Admittedly, one time, a British sailor was court-martialed, found guilty, and strangled for show. But usually, sailors could be confident their felonies would be chalked up to an accident or simple misunderstanding.

\* \* \*

While America had managed to benefit from the First Opium War without their firing a shot, it would not be so lucky during the Second Opium War that erupted between 1856 and 1860 as a result of unintended consequences of the first one.

After Great Britain cut its favorable deals with China at the end of that first war in 1842, America wanted to be given similar if not equal terms, and President Tyler sent over a lawyer named Caleb Cushing to get them.[11] The treaty he negotiated gave the United States virtually the same trade guarantees as the British, and added some key provisions—particularly the right to build churches and hospitals in any of the treaty ports and, at last, the official abolishment of a law that forbade foreigners to learn Chinese.

Meanwhile, the British had continued to push for their own additional concessions, even as China struggled to make them adhere to the terms of the first. In October 1856, again after a perceived slight to the British flag, a battle broke out, followed by a series of other skirmishes. Great Britain eventually seized Canton and China agreed to the Treaties of Tianjin, which broke down any significant remaining trade barriers. Westerners were given access to more ports, the right to settle in Beijing, and full legalizing of opium imports. (What the Chinese did after it arrived was their problem.) But, in 1860, after China proved slow to implement all of the terms of these treaties, British forces, with help from France and the United States, marched inland to Beijing and burned the emperor's summer palace. This led

---

11    While he wasn't, believe it or not, a direct descendent of John Perkins Cushing, they were both part of the same extended Cushing family of Hingham, Massachusetts, and definitely cut from the same cloth.

to the Beijing Convention—essentially a ratification of the Treaties of Tianjin.

Even though the United States never formally allied with France and Great Britain—in fact, at one point they pledged neutrality to the Chinese government—they ended up receiving most of the same concessions as the other two countries.

Altogether, the Beijing Convention and a series of subsequent accords are known fittingly in China as the "Unequal Treaties." While the Qing dynasty staggered along for the next fifty years, followed by a series of weak central governments, China wasn't strong enough to achieve international leadership again for another century, when Mao Zedong and the Communist Party took control.

The Second Opium War and the treaties that followed set the stage for the resentment, suspicion, self-righteousness, and frequent outright enmity that have marked the relationship between China and the West ever since—from America's racist Chinese Exclusion Acts of the late nineteenth century to the fact that the fentanyl crisis is often unfairly blamed on China's allegedly lax drug controls today.

\* \* \*

At the end of his life, Thomas Perkins was eulogized as "one of the noblest specimens of humanity to which our city has ever given birth."[12]

By the end of *his* life, John Perkins Cushing had founded the city of Belmont and was known as one of the most generous men in Boston.

---

12   Bebinger, "How Profits from Opium Shaped 19th-Century Boston."

By the end of *his* life, Warren Delano II had a sixteen-year-old grandson who would become president of the United States.

And, by the end of her life, their contemporary, the legendary Empress Dowager Cixi (1835–1908), who is considered one of the most powerful women in Chinese history, "was a regular evening smoker who advocated moderate use by retirees."[13]

In the end, the China trade, and the wars that resulted from it, is a cautionary tale about drugs, money, power, and greed—and the hypocrisy that inevitably hovers in the territory in between.

---

13  Chapman, "Taking Business to the Tiger's Gate."

# Chapter 22

# Americans Try Growing Their Own

When ancient Egyptians began to grow their own opium poppies, it reduced their dependence on Middle Eastern opium. Two thousand years later, when India's Mughal Empire encouraged farmers to start growing opium poppies, it did the same. During the nineteenth century, when farmers in American made serious attempts to grow their own, they too were trying to free themselves from dependence on foreign opium.

Americans had first tried to create a domestic supply during the Revolutionary War, when supplies of the painkiller were cut off by Great Britain. In fact, injured soldiers were treated with homegrown opium from Charlestown, New Hampshire.[1] Even Thomas Jefferson may have been involved. When not working on military and political strategies in the Virginia state government he did his patriotic duty by growing opium poppies in his gardens, poppies that were still cultivated at Monticello until 1992, when they were pulled up and destroyed along with all seed packets of "Jefferson's Monticello Poppies."[2]

Contemporary writings after the Civil War point to growing interest

---

1  Shattuck, *Green Mountain Opium Eaters*, 17–18.
2  There had been a drug bust at the University of Virginia (which was literally out the back door) and, spooked by the potential for bad publicity, the Board felt they had to take action. Jefferson's invaluable poppies went from the endangered species list to extinction in a matter of hours. See "Thomas Jefferson's Monticello," and Raver, "Poppies."

in cultivation in America. In a special feature in 1810 of the *American New Dispensatory*—one of the few reliable sources of information about drugs for doctors at the time—James Thacher calculated that an acre of rich, well-cultivated poppies could produce fifteen to twenty pounds of opium and that the grower could earn about $70 per acre from it, making opium more profitable than most grains. He added that while it was a lot of work to collect the poppy juice, fortunately "by far the greatest part of the whole labour of the season, may be performed by women and children." In conclusion, Thacher argued, "Every effort, therefore, to effect an object so truly interesting and important ought to be duly encouraged and rewarded."[3]

Beyond the money to be made, Thacher's rationale for why Americans ought to focus on cultivating homegrown poppies was his concern that foreign supplies might be adulterated. He should have been equally concerned about domestic supplies. For example, during the 1860s, an enterprising Vermonter named Welcome C. Wilson claimed to be able to manufacture more than 100 pounds of opium per acre of poppies. He tried to leverage his success by selling his spectacular seeds and custom processing equipment. The scientists to whom he sent the samples for testing, however, determined that the opium he claimed came from his special poppies was laced with a good deal of already-processed Turkish opium.[4]

He wasn't the only huckster making hyperbolic claims. Before Wilson's downfall, one of main competitors was fellow Vermonter Dr. Jonathan Moore, who used the opium from his garden to formulate Dr. Moore's Essence of Life, an elixir recommended for "consumption,

---

3   Thacher, *The American New Dispensatory*, 457.
4   Even now, in a good year, Afghani growers are happy if they can eke out a mere 25 pounds per acre. https://reliefweb.int/report/afghanistan/afghanistan-opium-survey-2017-cultivation-and-production.

difficult breathing, quinsy, spitting of blood, flatulence, fits, and hypo-chondriac afflictions."[5]

\* \* \*

One particularly interesting cultivation experiment in America took place at an unlikely place: the settlement in Mount Lebanon, New York, of the United Society of Believers, better known as the Shakers, a group known for their hard work, extraordinary craftsmanship, and religious devotion as well as their practices of communal living, celib-acy, and pacifism. They believed in racial and gender equality and, most of all, giving their "hands to work and hearts to God." Gardening was an integral part of Shaker life and faith. They believed they had a mission to transform the soil from "rugged barrenness into smiling fertility and beauty."[6] As it turned out, they were opium dealers.

The Shakers planted a total of 200 acres of medicinal plants in their communities with 50 at Mount Lebanon alone and processed almost 300 varieties of indigenous and imported natural ingredients.[7] While famous for their seeds (which they were among the first to sell in printed paper packets), the Shakers earned as much if not more selling what passed for medicine at the time.[8] Many of these remedies were single-plant extracts of everything from aconite to yellow bark. Others were cures for coughs, insomnia, asthma, even coloring gray hair. "Gray hair," their label read, "may be honorable, but the natural color is preferable." They also produced endless remedies for distressed

---

5 Chaplan, *Urban Treasure Hunter,* 153.
6 Andrews and Andrews, *Work and Worship Among the Shakers,* 48.
7 There were more than two dozen Shaker communities founded in the 1800s, primarily in the northeast.
8 Buchanan, *The Shaker Herb and Garden Book,* 44.

bowels, which have evidently been the bane of humanity since ancient Sumeria—and for which naturally constipating opium would have been an excellent drug. But they saved their most exuberant claims—and their opium—for an elixir called "Pain King."

In a country overflowing with tinctures and panaceas, the Shakers claimed the high ground with Pain King, the "Absolute Monarch of Distress and Suffering," "The King of All Great Pain Destroyers." Pain King provided "instant" relief for pain associated with sprains, burns, wounds, toothache, neuralgia, sore throat, diarrhea, and even diphtheria. According to one customer, "It not only gives relief, but it cures." That customer was so enthusiastic, he asked for a double order, perhaps figuring that, like drug users throughout history, he could sell some to cover his costs and feed his own habit for free.[9]

Reassuringly for such a high-powered product, its label said it was "perfectly safe," and gave instructions how to administer it for different conditions: gargle and apply externally to treat a sore throat; hold half a teaspoonful in your mouth for a toothache; and lie "on a wet cloth covered with cotton flannel for a backache." Even if it would not pass today's FDA guidelines, the Shakers followed a very precise formula that reflects careful research into the most effective ways to prepare the drug in order to benefit from its "active principle" along with other ingredients that they thought were complementary. They describe the formula as:

20 gal. water. 10 lbs. Witch Hazel bark stir every-day for one week. 20 gall. strong alcohol—oils Spruce Sassafras Peppermint Camphor gum, dissolve in separate portions of alcohol and mix

---

9   White, *Shaker Almanac, 1884*, 26.

altogether—Opium, reduce to a miscle condition with warm water and Masher till all parts are accessible to oil and water, after the former mixture has been put together and well stirred. Or find out by Dispensatory, or by *The Pharmaceutical Journal* what is the proper strength of Alcohol to extract the active principle of Opium and preceed accordingly then mix with the rest.[10] [*sic*]

The crop was an important part of the Shakers' culture as well as finances. Shaker women collected the sap the same way Middle Eastern farmers had been doing for centuries. In 1906 Sister Marsha Bullard waxed eloquently about her memories of white-capped sisters "stooping among the blossoms to slit those pods from which the petals had just fallen. Again, after sundown they came out with little knives to scrape off the dried juice." Sister Marsha also observed that its production was one of "the most lucrative as well as the most picturesque of our industries."[11]

If their consciences needed further balm, the Shakers could take some refuge in the fact that they filled a crucial need during the Civil War. Since they were pacifists, they petitioned President Lincoln to exempt them from serving in the military, which he granted reluctantly, grumbling that, given their strong character, they "ought to be made to fight [as] we need regiments of just such men as you."[12] Although they demurred, the Mount Lebanon Shakers did set aside ten acres of poppies so they could supply opium for military doctors.

While, in retrospect their involvement in the opium business may seem out of character, for the Shakers and their contemporaries, it

---

10    Miller, *Shaker Herbs.*
11    Jensen, *With These Hands*, 56.
12    Hancock, "FAQs—Hancock Shaker Village."

was just another ingredient, one of hundreds of the medicinal flowers and herbs they grew. While they were among the few Americans who found ways to grow poppies with a high enough morphine content to make a successful product, their crop still represented an insignificant percentage of the 50,000 pounds of Middle Eastern opium sold annually in New York City alone—an indication of just how lucrative domestic cultivation could have been.[13]

\* \* \*

Since opium had proven so effective in treating wounded soldiers in the Civil War, farmers continued to explore domestic cultivation— even after early reports of addiction among veterans began to appear. *Scientific American*, that most respected of science magazines, didn't want earlier failures to discourage American entrepreneurs from trying to produce opium, and regularly wrote about ongoing experiments. The June 5, 1869, issue argued that poppy farming surely could succeed once farmers found the right soil and climate.[14] One reporter said he'd seen it growing "spontaneously on every uncultivated spot" in Ohio and that, in Texas, "acres of poppies stood as thick as wheat in a wheat field; and yielded excellent opium" (July 3, 1869). Another issue that year remarked on the fact that the *American Journal of Pharmacy* had tested some laudanum made from Virginia opium and found it as good as any from Turkey, although they were fairly dismissive of attempts in Vermont.

But for a variety of reasons, American farmers never achieved sufficient economies of scale to attract enough investors to make opium

---

13    "Scientific American Archives," August 21, 1852.
14    "Scientific American Archives."

poppies a major cash crop. Even after a farmer in Minnesota named Emil Weschcke grew *Papaver somniferum* that yielded a more than respectable 15 percent morphine, the Department of Pharmacy at the University of Wisconsin didn't believe it was possible to grow it efficiently enough to make it viable.[15] By the end of the Century, *Scientific American* finally threw in the towel, placing the blame squarely on the fact that there wasn't enough cheap labor (March 5, 1898). They were undoubtedly right. After all, even the much-admired Shakers had relied on child labor for the labor-intensive harvesting of the poppies, and in the late 1800s the increasingly powerful labor movement was finally beginning to make that source of labor a thing of the past.

---

15   Which is odd, since that compares favorably to Turkish opium.

# Chapter 23

# Good Intentions, Tragic Results

In the nineteenth century, three relatively unknown scientists transformed the use and abuse of opium: Friedrich Wilhelm Adam Sertürner, Alexander Wood, and C. R. Alder Wright—the developers, respectively, of morphine, the hypodermic syringe, and heroin. All three believed they had developed a more effective and *less* addictive way to use opium to relieve pain. All three were wrong. And two of them learned the hard way.

Sertürner (1783–1841) made arguably the most important chemical discovery of the nineteenth century: how to isolate an individual active agent from a plant.[1] It may sound simple, but while brilliant physicians had been developing remedies based on *combinations* of plant extracts since before Hippocrates, dosages were still hit-and-miss at best—especially in the case of poppies and other natural psychotropics because of the different amounts of active ingredients in each plant. After Sertürner's discovery, however, it was no longer necessary to try to cultivate poppies with similar morphine content, which, as

---

1  Around the same time, a French chemist named François Derosne produced a different alkaloid from opium sap called "narcotine." He thought he had found the active agent of opium but, while it has some cough-suppressing properties, they are far less than those of morphine. Similarly, a French chemist named Armand Seguin announced that he had developed a process for isolating alkaloids from opium. But he didn't publish his results for another decade.

Avicenna learned so many years before, was a thankless task. Now, a measurable amount of a specific chemical, in this case morphine, could be extracted from every plant that contained it.[2] In many ways, this achievement marked the birth of modern pharmacology.[3]

For someone who made such a major scientific discovery, Sertürner had a remarkably humble background. He was not a famous professor—in fact he did not even go to college. He didn't have access to advanced lab equipment—only a few basic tools, some of which he fashioned himself. He wasn't even a professional chemist— just an apprentice at an apothecary, a job he got out of desperation at sixteen when his parents and their patron, an Austrian prince, all died in 1798.

What Sertürner *did* have was an abundance of curiosity and intuition. He spent every moment he could spare experimenting with compounds: heating, cooling, precipitating, hydrating, oxidizing, compounding, shaking, rattling, and rolling different chemicals to see what new compounds he could come up with. Each time he discovered a new one with promise, he used the most advanced analytical tools available to study it—i.e., he took some himself.

One time, conveniently, Sertürner did an experiment with poppy juice when suffering from an excruciating toothache. (Around the same time, Thomas De Quincey, the infamous author of *Confessions of an English Opium-Eater*—the classic of addiction literature—used a

---

2  Nowadays, chemists change the actual molecular structure of existing *compounds*...Sertürner took the first step, i.e., figuring out how to isolate individual organic chemicals.

3  One reason no one had done it before is that scientists had been laboring under a profoundly inaccurate assumption: that active elements had to be acidic. This assumption makes sense since we think of "acids" as causing stronger reactions than "alkalines" (i.e., compounds with a pH over 7 percent). Sertürner discovered the error of their ways...in fact, he's the one who gave these compounds the name *alkaloid*.

similar excuse.) Sertürner had just performed a chlorate precipitation, which means he mixed some poppy juice with sodium chlorate and waited to see what happened next. It turned out he had created a new "salt," which he sampled. As his tooth pain began to fade, he fell asleep and he remained asleep for several hours. Impressed, he persuaded a few young friends to experiment with the salt along with him so he could evaluate dosages. They determined empirically that 15 mg was just about right.[4]

The conventional wisdom is that Sertürner chose the name "morphine" as an homage to Morpheus, the Greek god of sleep. Others, however, claim that the name comes from the word *morphe*, meaning "shape" as in the "shape of dreams."[5] Whether it's a reference to a Greek god or something as fundamental to our psychology as the dreams that opium evokes, the name speaks to the power of the drug, which Sertürner was on the treacherous road to experiencing firsthand. Within a few years of sampling the new "salt," he realized how addictive it was and, unlike some drug company executives today, didn't hide the fact. In 1812 he wrote, "I consider it my duty to attract attention to the terrible effects of this new substance I called *morphium* in order that calamity may be averted."[6]

But for better *and* for worse, the genie that Sertürner had let out would never go back in the bottle.

The scientific establishment eventually realized the magnitude of his achievement: he received an honorary degree from Jena University in 1817, followed by awards from universities in Marburg, St. Petersburg, Batavia, Paris, and Lisbon. In 1831, he won 2,000 francs

---

4   This could be considered the Goldilocks clinical-trial protocol.
5   Dormandy, *Opium*, 116.
6   Quoted in Dormandy, 117.

from the Institute of France and the title "Benefactor of humanity,"[7] a sentiment undoubtedly shared by the millions who have found relief from his discovery of the drug, although likely troubling for those who have lost loved ones to it.

Later in his career, Sertürner gave up on psychopharmacology and moved on to more reliably lethal weapons, making several major improvements to the design of firearms and bullets. As he got older, he became increasingly depressed and increasingly addicted to his creation. He died in 1841, at age fifty-seven. While the official cause of death was "dropsy" (likely congestive heart failure), after he died, his doctor said that he had begun to exhibit "successive, aggravated hypochondriacal alterations in his frame of mind and obvious quiet disturbances of mood." While his morphine use must have taken a toll, the symptoms actually sound more like those of a dying person who was *under*medicated. His tolerance may have been so high by then that he wasn't able to get a clinically effective amount into his system.[8]

* * *

Sertürner had solved one half of the dosage equation—how to create "replicable" doses of an alkaloid. Several decades later, two other scientists—Charles Gabriel Pravaz and Alexander Wood—solved the other half of the equation: the most efficient way to deliver those doses.

---

7   Aggrawal, *Narcotic Drugs*. Also see Krishnamurti and Rao, "The Isolation of Morphine by Sertürner."

8   Schmitz, "Friedrich Wilhelm Sertürner and the Discovery of Morphine." Twenty years later, E. Merck & Company actually began manufacturing morphine. Although the question of who can use the name "Merck" is making many lawyers rich, this is, indeed, the forerunner of the modern-day Merck & Co. and publisher of the famous *Merck Manual*. It began as a German drugstore in 1668. The company does not currently manufacture any prescription opioids.

The latex (also called "poppy tears") that oozes from the opium poppy has been harvested for thousands of years by making shallow incisions during the afternoon. The juice dries overnight and the thick black or brown opium is scraped off the next morning. It contains 10–15 percent morphine, 1–3 percent codeine, as well as thebaine (1–2 percent) from which oxycodone is made.

Woven grass baskets with poppy seed pods and other ceremonial items such as flowers and gemstones were found at the Cueva de los Mer-ciélagos ("Bat Cave"), a Neolithic site in southern Spain. DNA testing showed traces of opium use in the bones of one of the skeletons.

The Ebers Papyrus (1875 repro-duction), one of the most famous papyruses from ancient Egypt, includes formulas that use various parts of the poppy to treat head-aches, constipation, sore muscles, and the flu.

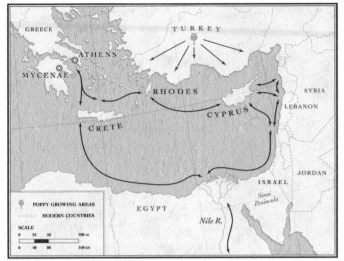

Ancient merchants brought opium from the interior of modern-day Turkey and Iran to cities all along the Mediterranean Coast where it was used in religious rituals as well as medicinally.

Some of the earliest representations of the poppy can be found in the bas-reliefs of temples to the Greek deities, such as Demeter, the goddess of agriculture, who is shown here carrying a fistful of poppy heads.

Opium was used as an anesthetic in the Middle Ages, in combination with other classic sedating ingredients such as hemlock, mandrake, and henbane. Here, a surgeon's assistant, often a nun, soaks a sponge in opium and places it over a patient's nose as needed—similar to how ether was used in the late 1800s.

This 1860s tobacco label shows a Native American presenting Christopher Columbus with a sheaf of tobacco leaves. His sailors took to the habit quickly and spread it to the Far East on subsequent voyages where they began blending it with small balls of opium.

In 1602, Queen Elizabeth wrote this letter to Chinese Emperor Shen Zong and gave it to George Waymouth who hoped to find a Northwest Passage to the East. She wanted to negotiate directly with the Chinese in order to give the British East India Company a competitive edge over the Portuguese and Dutch. Waymouth turned back after a near-mutiny in the Hudson Strait.

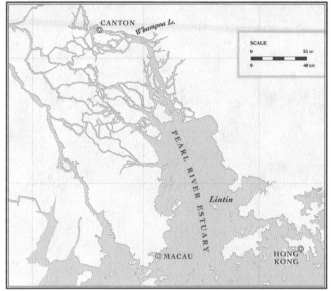

China began to allow foreign merchants to do business on the mainland during the 1500s but restricted trade to the Pearl River Estuary. When Commissioner Lin introduced opium bans in the 1800s, the many hidden coves and harbors along the coast provided ample opportunities for smuggling opium and other contraband and contributed to the breakout of the Opium Wars.

This portrait of Canton in the 1800s—owned by Boston's Museum of Fine Arts, which was funded in large-part by American opium traders—reflects the West's romanticized image of what was actually a bustling commercial harbor crowded with small junks and sampans. The headquarters of the Western merchants were clustered along the river, outside the city walls.

By the late 1700s, the British had taken control of the main opium growing regions of India and were using (and abusing) locals to plant, harvest, and process the opium, as pictured in this Indian factory. About fifty 2.5–3.5 pound balls would be packed into chests and sold at auctions in Calcutta, much of it to be smuggled into China.

Commissioner Lin arrived in Canton in 1839 determined to end the opium trade and launched a crackdown that provoked the First Opium War. Lin remains a hero to Chinese people all over the world and this statue in Manhattan's Chinatown is a reminder of the West's complicity in the addiction of millions of Chinese during the 19th century.

The Shakers of New Lebanon, New York, grew and processed opium poppies. This advertisement from the 1884 Shaker Almanac sings the praises of "Pain King," one of their opium-infused products.

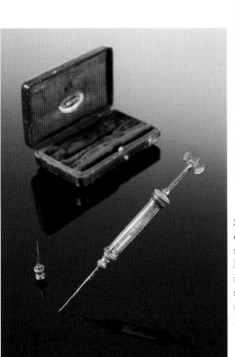

Several doctors took credit for the invention of the hypodermic needle. This late 1800s syringe, based on the design by Doctor Charles Pravaz, was made by a surgical instrument maker in Paris. By the late 19th century, syringes in pocket-sized cases with extra needles and vials were available in the Sears, Roebuck catalog for as little as $2.00.

With the surging popularity of smoking, opium use became increasingly common in 19th century America. In his 1883 book, The Living Death, Dr. H. H. Kane reported that opium use was prevalent throughout American society, including by "gentlemen of leisure" and "ladies of good families."

THE QUEEN OF CHINATOWN
BY JOSEPH JARROW

HURLED BY HIGHBINDERS THROUGH THE RAT PIT'S DOUBLE TRAP.

In the late 1800s Chinese Americans became the first of a succession of immigrant groups blamed for America's drug problems. In this Broadway melodrama, a man tries to "rescue" upstanding ladies who have been lured into a Chinese "den of iniquity," only to be thwarted by a Chinese assassin.

"There are 100,000 total marijuana smokers in the US, and most are Negroes, Hispanics, Filipinos, and entertainers. Their Satanic music, jazz and swing, result from marijuana usage. This marijuana causes white women to seek sexual relations with Negroes, entertainers, and any others."

—Harry J. Anslinger

Director of the Federal Bureau of Narcotics from 1930 to 1962, Harry Anslinger remains a controversial figure who has been criticized for sensationalizing the evils of drug use and exaggerating his successes while establishing race-based policies of drug enforcement that continue to this day.

As a famous African American, Billie Holiday was a target of the Federal Bureau of Narcotics' in the 1940s and '50s. One of the most talented jazz singers in American history, she was eventually arrested and handcuffed to her death bed in New York's Metropolitan Hospital for drug possession.

The United States military argues that patrols such as these are necessary in order to take control of poppy-growing districts of Afghanistan without alienating the local population. In 2017, opium poppy cultivation in the country again reached record highs.

After telling President Nixon that he was "on your side" in a 1970 visit to the White House, Elvis was given a badge from the Federal Bureau of Narcotics. According to Priscilla Presley, he thought the badge would allow him to take drugs (and guns) into any country without fear of confiscation. At the time, he relied heavily on amphetamines and barbiturates.

Safe (or supervised) injection sites are a controversial harm-reduction strategy. They protect users from disease and overdose by allowing them to take their drugs in a place where clean needles and naloxone are available. At some sites, batch testing and counseling are also offered.

When a person eats, drinks, smokes, or even snorts a drug (the only forms of ingestion prior to Pravaz and Wood), the effect depends on the characteristics of an individual's digestive or respiratory systems.[9] The hypodermic syringe changed that, by making it possible to inject a drug directly and reliably into the bloodstream.

Scientists were well aware that subcutaneous injection had the potential to cause or treat pain instantly. There were examples throughout nature: bees, poisonous snakes, and spiders.[10] Warriors had used the same principle for centuries when shooting poison darts. But, until the 1800s, no one had found a way to overcome the many challenges involved in getting a measurable amount of a drug directly under the skin and into the bloodstream without spillage, dilution, or contamination.

In ancient Egypt, a pump was used to force liquid into the body to embalm the dead and give enemas to the living. The ancient Roman physician Galen tried opening veins and pouring drugs into them. Crude subcutaneous injection had begun taking place by the Renaissance. In 1656, a group of curious and mischievous young British intellectuals gathered in the lab of an Irish aristocrat named Robert Boyle, who is considered the father of modern chemistry. One of the other men was Christopher Wren, who would one day become a famous scientist and architect. Wren filled an animal bladder with opium, attached a goose quill, and, in what can only be described as

---

9   Morphine pills can be powdered and snorted. And there's at least one report of someone dying from smoking a morphine patch. See "Morphine Patch Prison Death 'Accidental'"; and "Ask Erowid: Can Morphine Pills Be Smoked?"

10   The first officially recorded morphine murder was committed in the early 1820s by a French doctor named Edme Castaing, who allegedly added morphine to his friend Auguste Ballet's wine shortly after Auguste had made Castaing the sole beneficiary of his will. Although he claimed to be innocent to the last, he was executed by hanging in December 1823. See Aggrawal, "The Story of Opium."

PETA's worst nightmare, administered opium to a dog that he and his co-conspirators had tied to the four legs of a table.

We don't know what the procedure felt like for the animal who'd given up his or her bladder to science, but we do know what it was like for the dog. After being untied, it "began to nod with his head and faulter and reel in his Pace [sic], and presently after appeared to be stupefied." What happened next is sadly poignant and foreshadows the countless human overdoses and near-overdoses that intravenous opioid injections have caused in subsequent years. While several in the group began to make bets on whether the dog would survive, Boyle took him outside and "whipped [him] up and down a neighbouring Garden," forcing him to stay awake "until he had successfully walked it off." [11]

In the early 1800s, an Irish doctor named Francis Rynd invented a hollow needle that could puncture the skin without causing bleeding. He, however, only used gravity to deliver morphine (essentially "pouring" it down a tube connected to the hollow needle). Other doctors tried equally crude methods such as making incisions or using large needles to make holes in the skin and push in a morphine pellet; or blister the skin, peel it back, and apply a poultice with the drug. The results were still unsatisfactory. What was needed was a way that combined those two technologies in order to force a precise amount of the drug into the bloodstream through a tiny puncture. [12]

Credit for the development of this type of precision syringe/needle has been given to both Charles Pravaz as well as Alexander Wood, often depending on whether the creditor is an Anglophile or Francophile.

---

11  Dorrington and Poole, "The First Intravenous Anaesthetic." Also see "The History of Injecting, and the Development of the Syringe."
12  Pates, McBride, and Arnold, *Injecting Illicit Drugs*, Ch. 1.

Pravaz's claim is based on illustrations of syringes that are named after him. They are things of beauty, made of silver and packaged in fitted silk and velvet-lined cases, an indication of how Victorian men and women who needed to find relief (and, subsequently, maintain their habits) were addicted to the aesthetic as well as the anesthetic properties of getting their fix.[13]

Yet while Pravaz was a brilliant orthopedist who developed treatments for scoliosis and clubfoot, his claim to being the inventor of the hypodermic syringe was weaker than that of Wood, who was the first and most influential person to describe subcutaneous injection. His life is also a far more relevant tale because it seems that, even before he created his first syringe, his wife Rebecca had developed a lethal fondness for morphine. As he worked on his contraption she doubled down by serving as a beta tester. The rumor is that she became the first recorded victim of an overdose after a hypodermic injection. Of course, when the cause of death is a drug overdose, the "usual suspects" are always family members (who usually have a motive) and the family doctor (who usually has the means). Seeing as how Wood was both, it's natural, if perhaps unfair, to wonder whether the overdose was accidental.[14]

In terms of the history of opium, Wood's invention is also more significant since he used his hypodermic to treat pain, whereas Pravaz used his to treat an aneurysm. However, like most scientists researching the use of hypodermics for morphine at the time, they shared an incorrect assumption: that the poppy's addictive nature was due not

---

13  Phisick Medical Antiques, "Pravaz Hypodermic Syringe in Silver." (Note, the tubes were soon made of glass instead.)

14  Some have challenged the veracity of these rumors, for instance, Davenport-Hines, *The Pursuit of Oblivion*, 100–101. He says Rebecca survived her husband, living until 1894.

to the morphine itself but the route of delivery. In other words, they assumed that when someone has a "hunger" for something, it's triggered by a sensation in the stomach. If they could bypass the stomach by delivering the drug directly into the bloodstream, they thought they could bypass this addiction mechanism.

Most doctors of the time made another incorrect leap of faith. They thought the primary action of injection was local. When they shot morphine into the part of the body where the pain was, it took care of the pain. They assumed that the patients' overall sense of well-being simply came about from relief of that pain. Considering that even at that time, habitual recreational users enjoyed the drug regardless where they stuck the needle, it's remarkable how long some doctors clung to that assumption.

Later in the 1800s, an English chemist named C. R. Alder Wright figured out how to further refine Sertürner's morphine isolate by tinkering with its actual molecular formula, thereby crossing the line from "natural" to semisynthetics.[15] In other words, he wasn't just isolating a specific alkaloid in poppy juice, he was tinkering with the alkaloid itself. His breakthrough, which he didn't appreciate at the time, came when he mixed up various existing drugs with other compounds known as acetyls. These compounds tend to cause rapid reactions, resulting in drugs that are, if correctly mixed, more pure, powerful, effective, and/or less likely to cause side effects. One day, he boiled morphine with twice its weight of acetic acid. After performing a few other chemical tricks of the trade, he ended up with crystals

---

15  Semisynthetics are created by performing experiments on a natural isolate of a plant, such as morphine. True synthetics are created by taking random molecules (often from petrochemicals) and combining them in ways whose effects are, sometimes, as much of a surprise to the scientists who discover them as the people who eventually take them. Once a scientist knows the molecular formula for a natural or semisynthetic drug he can make a "pure" synthetic.

of several substances, including "diacetylmorphine." He did a similar process with codeine and administered it subcutaneously into dogs and rabbits. The results were similar in each case: the animals' pupils dilated, they had the dry heaves, their breathing quickened and then slackened, their heartbeat slowed, their coordination worsened, and their rectal temperature got lower. Fortunately, the animals recovered from these symptoms, although it usually took twenty-four hours.[16] Because Wright was largely driven by curiosity to discover more chemicals with interesting properties rather than an entrepreneur's commitment to monetizing them, he decided to move on to other things.

Wright might have been forgotten if not for a chemist at Bayer named Heinrich Dreser who, two decades later, continued where Wright left off and, with the help of one of his colleagues, soon recognized diacetylmorphine's amazing pain-relieving and cough-suppressing power. They gave it the name *heroiche,* a German word meaning "heroic" or "strong," because it was so much stronger than morphine. Instead of chalking it up to just another interesting chemical interaction as Wright did, they easily persuaded Bayer to begin producing it in serious quantities and launch a marketing campaign that even involved mailing out free samples.[17] Soon, you could buy heroin tinctures wherever fine drugs were sold—and not just for respiratory problems but also for headaches, colds, and those notorious "female problems." It was a drug for the rich.

Doctors were so impressed with heroin's purity, they thought they could use it to treat morphine addiction, just as the morphine

---

16  Wright and Beckett, "On the Action of Organic Acids and Their Anydrides on the Natural Alkaloids. Part I."
17  "Heroin, Prescribed for Coughs and Headaches, was a Trademarked Medicine Produced by Bayer Company."

developers before them had thought they could use *it* to treat opium addiction. Dreser, like so many of opium's innovators, would pay the ultimate price for this misguided assumption.[18] While he became rich from his discovery of heroin (and, later, aspirin) he also became an addict and died in 1924.

The quest for a nonaddictive opioid—or safer way to deliver it—continued, as it continues today. So far, the resulting drugs have proven to be increasingly addictive and deadly.

---

18  He unfairly took the sole credit for aspirin, ignoring the contribution of his supervisor Hoffman as well as a Jewish chemist named Eichengruen. They received neither the money nor the credit they deserved—a fact Eichengruen pointed out in the most scathing terms after World War II when he emerged from a concentration camp and wrote a denunciation of Dreser's "discovery" of aspirin. See Askwith, "Heroin, Bayer and Heinrich Dreser."

# Chapter 24

# The Agony and the Ecstasy

Of all the remedies it has pleased almighty God to give man to relieve his suffering, none is so universal and so efficacious as opium.[1]

—Thomas Sydenham (1624–1689)

Nobody will laugh long who deals much with opium: Its pleasures even are of a grave and solemn complexion.[2]

—Thomas De Quincey (1785–1859)

Conceive whatever is most wretched, helpless, and hopeless, and you will form as tolerable a notion of my state, as it is possible for a good man to have. In the one crime of opium, what crime

---

1   While Paracelsus created the first *laudanum* formula (at least in the West), Sydenham, who lived 150 years later, was the first to market one.

2   All De Quincey quotations in this chapter are from his *Confessions of an English Opium-Eater*. He originally wrote a version for a literary journal when he was young and in desperate need of money. This is the expanded version he wrote later in life. It was included in a collection of some of his writings with an introduction and notes by George Armstrong Wauchope.

have I not made myself guilty of!—Ingratitude
to my Maker! and to my benefactors—injustice!
and unnatural cruelty to my poor children!—
self-contempt for my repeated promise—breach,
nay, too often, actual falsehood! After my death,
I earnestly entreat, that a full and unqualified
narration of my wretchedness, and of its guilty
cause, may be made public, that at least, some
little good may be effected by the direful example...[3]

—Samuel Taylor Coleridge (1772–1834)

Many people believe the opioid crisis in America began at the turn
of the century. Though they are correct, most are probably thinking
of the wrong century.

By 1800, enterprising doctors and pharmacists were already enthusiastically promoting the pain-relieving properties of opium, seemingly
with little regard for its potential for abuse—much in the same way
that doctors and pharmacists in years to come would, in turn, enthusiastically promote morphine, heroin, and prescription opioids. So
prevalent was its use that there were in fact thousands of opium-based
medicines in the nineteenth and early twentieth centuries.[4] For 200
years, the most famous of all was Dover's Powder, the brainchild of
Thomas Dover. It included one ounce each of opium, licorice, and

---

3  Coleridge et al., *Letters of Samuel Taylor Coleridge.* Letter to Josiah Wade, June
   26, 1814.
4  "In the middle of the nineteenth century, there were some seventy thousand
   remedies with secret formulas in the drugstores of Europe and the Americas
   (the "Tonic of Dr. X," the "Miraculous Water of Z," etc.), which almost invari-
   ably used psychoactive drugs and were advertised by all kinds of publications,
   billboards, and posters." See Escohotado, *A Brief History of Drugs,* Ch. 10.

ipecacuanha, along with four ounces of saltpeter and vitriolated tartar (i.e., treated with sulfuric acid). The patient was instructed to put it in a glass of white wine and take it at bedtime. Dover, who, among other achievements, was commander of the vessel that rescued Alexander Selkirk (the actual Robinson Crusoe), was well aware of its potency and the risk of overdose. Cautioning that it be used as directed, he said: "Some apothecaries have desired their patients to make their wills before they venture upon so large a dose,"[5] although, like today's heroin dealers, Dover may have known that users would take his cautionary advice as a guarantee of its strength rather than a warning of its potential risk.

It was as natural in the West in the nineteenth century to have one or more opium-based nostrums in the medicine cabinet as it is today to have aspirin, cough syrup, or Pepto-Bismol. After all, opium in small, prescribed doses *can* cure headaches, coughs, and diarrhea. Most people in the 1800s thought nothing of taking a sip of laudanum to alleviate pain, get some relief after a back-breaking day of work, and deal with female complaints that male doctors considered baffling as well as equally psychosomatic male complaints that were considered noble. Those who weren't taking it for a physical complaint used it to enhance their creativity, explore new ways of thinking and feeling, or even just to have some risky recreational fun.

All that was needed to get started was a pharmacist's recommendation, a friend's urging, an enticing label, a desire to be rebellious,

---

5   Dover thought those pharmacists were being more than a little histrionic. Regardless, it did foreshadow Thomas De Quincey's experience of trying to get life insurance and being turned down by multiple companies due to his opium habit. De Quincey took more than a little wicked pleasure in the fact that the well-meaning adjuster, assigned to study whether opium did indeed lead to an early grave, died long before he did. See De Quincey, *Confessions of an English Opium-Eater*, 117ff.

or simply an adventurous nature. After that, it was one long straight road to hell—or so conventional wisdom of the time would have you believe. In Balzac's play *Comédie du diable,* opium was depicted literally as the road to hell. The Devil says that morphine was causing a population explosion in his realm.[6]

People have a natural tendency to believe they live in exceptional times—that, in the case of opium, for example, addiction wasn't as common, withdrawal wasn't as difficult, and overdoses weren't as frequent back then (whenever *then* was). There may be some truth to the notion that opium's dangers were less acute in the 1800s. However, easier access to strong Turkish opium, popularizations of its pleasures and perils by writers, the introduction of opium *smoking,* and the developments of morphine and hypodermic injection dissolved most of the differences between the dangers of opioid use back then and today. The scale of the problems today may differ, but the reasons people use opioids, the mechanisms of addiction, the risk of overdose, and the greed that drives dealers to adulterate their supply—all the problems that doctors, nurses, social workers, addiction specialists, and patients confront every day—have corollaries in the nineteenth century: every behavior, every opinion, and every prejudice.

There is something deceptively charming, almost romantic, about the photographs and illustrations from that time: an old apothecary with a smiling proprietor in front; a dreamy-eyed Chinese man of indeterminate age lying on his side, head heavily raised to a pipe; a delicate, suggestively hypochondriacal woman on a fainting couch; a debauched writer, composing poetic masterpieces, but destined for an early grave.

---

6   Cited by Dormandy, *Opium,* Ch. 16.

Then, there are the most misleading images of all—those hyperbolic "old-fashioned" labels promising to cure everything from a baby's colic to an adult hangover.

While we may realize that those words and images are clichéd—if not embroidered within an inch of an outright lie—they can make us smile. Certainly, we'd never be taken in by claims like that today, right?

Indeed, the labels are now valuable ephemera, collectibles much like cigarette magazine ads from the 1950s and psychedelic Timothy Leary posters from the 1960s. As one modern writer—and former addict—described the influence of those 1800s labels: "What troubles me about this kind of titillating drug pornography is how it neatly side-steps any contentious issues, fetishizes opium paraphernalia, [and] avoids mention of withdrawal and addiction."[7]

These images give us a distorted view of the past—they make us forget that real people experienced the real agony and ecstasy of opioids back then, too.

Then, as now, much of the conventional wisdom about opioid use was based on a tangled web of narrow sociological studies; sensationalized news stories; bad, inaccessible, biased, or conflicting data; ethnic prejudices; fictional portraits; and a romanticizing of creative and spiritual struggle—all of which fed stereotypes that continue to affect public perception and legislative policy today.

---

7   Keeling, "The Drugs Don't Work."

## MYTH #1:

### OPIUM *SMOKING* IS A VICE OF THE POOR.

Then, as now, Americans have been able to rationalize or outright ignore the effect of certain drugs on society by pretending that only the poor are affected.[8]

In the early 1880s, for example, a doctor named H. H. Kane decided to test this assumption by making a careful study of opium use. He wrote: "Of what class are those who smoke opium? The answer is, representatives of all classes—merchants, actors, gentlemen of leisure, sporting men, telegraph operators, mechanics, ladies of good families, actresses, prostitutes, married women, and single girls."[9] He claimed that opium smoking was so common that it functioned as almost a social lubricant the way one might think of drinks at a cocktail party or around a keg of beer today. "Men and women," he wrote, "often meet to smoke, talk, and enjoy that state that comes as near as it is possible for an American to come to the *dolce far niente* of the Italian."[10]

He also pointed out that while many people did consider smoking a social event—others preferred to smoke alone, even using the time to catch up on correspondence or listen to music.[11] (Of course, if they

---

8  For example, in twentieth-century America, heroin was considered a vice of the poor so the country didn't respond with any urgency to the widespread use of the drug until, as we'll see, "everyday" people began to shoot heroin as a replacement for prescription painkillers.

9  Kane, *Opium-Smoking in America and China*, 72–73.

10  Ibid., 43.

11  Nineteenth-century world leaders weren't above fortifying themselves with opium either. Germany's "Iron Chancellor," Otto von Bismarck, always dosed himself before important government meeting, as did England's William Gladstone before addressing the House of Commons. This foreshadows the often-overlooked truth that people on maintenance doses of opioids today are able to go about their daily business. See Jay, *Emperors of Dreams*, Ch. 2.

continued to take pipe after pipe they would indeed drift off into their personal opium dreams.)

MYTH #2:

THEN, AS NOW, IMMIGRANTS ARE PRIMARILY TO BLAME FOR THE NINETEENTH-CENTURY OPIOID CRISIS IN AMERICA.

Although the Chinese introduced opium *smoking* to America, the country was already awash in opium-based elixirs and injectable morphine by the time Chinese immigrants arrived in the mid-to-late 1800s looking for jobs working on the railroads.

Americans took to the habit quickly. Some wealthy smokers set up Oriental-style rooms in their homes and had friends over to smoke.[12] But stories about innocent young white girls being lured into dens of iniquity made better press than the secret vices of the rich. Still, even after the city fathers in San Francisco forbade white people from visiting Chinese opium dens, as much as 100,000 pounds of the drug arrived annually in San Francisco Harbor—a quantity that would have been improbable for the 14,000 Chinese men and women living in the city in 1900 to smoke by themselves.[13]

The myth that foreigners and immigrants are responsible for our drug crises continues today in the belief that much of America's opioid crisis is caused by heroin and stronger drugs from China that are smuggled in across the Mexican border.

---

12  Ibid.
13  "Scientific American Archives," March 5, 1898; "San Francisco History—Population." It's not inconceivable. A heavy user might smoke more than nine pounds of opium a year. See Kramer, "Speculations on the Nature and Pattern of Opium Smoking."

## MYTH #3:

### OPIUM IS WORSE THAN ALCOHOL.

Alcohol was used back then as it is today, by more people than opium and the results were far worse. "The opium-smoker," Dr. Kane pointed out, "does not break furniture, beat his wife, kill his fellow-men, reel through the streets disgracing himself or friends, or wind up a long debauch comatose in the gutter. He is not unfitted for work to the same extent that an inebriate is."[14]

## MYTH #4:

### VIRTUALLY ALL OPIUM ADDICTS LIVED IN MAJOR CITIES.

Then, as now, there were addicts all over the country. There's a whole book devoted to the history of early opium addiction in Vermont alone.[15]

## MYTH #5:

### WITHDRAWING FROM OPIUM THEN WAS EASIER THAN WITHDRAWING FROM OPIOIDS TODAY.

One recovering user described withdrawal in terms similar to those used by modern opioid users, saying, "The entire surface of the body was pricked by invisible needles....Every joint...was racked with consuming fire, while intermittently from every skin-pore there issued a deluge of sweat...the soul was oppressed with disquietude, the heart fluttered like a wounded bird, and the brain faltered from irresolution."[16]

---

14    Kane, *Opium-Smoking in America and China*, 74–75.
15    Shattuck, *Green Mountain Opium Eaters*.
16    William Cobbe, late 1800s, cited by Hodgson, *Opium*.

MYTH #6:

## PEOPLE WHO SMOKE OPIUM LIE ON THEIR SIDES BECAUSE THEY ARE SO IMMOBILIZED.

This is the most insidious myth of all—and another one that continues to this day—because it perpetuates the image of the degenerate opium smoker wasting his or her day lying on a squalid mattress or couch. A man named Steve Martin, who traveled to Southeast Asia to collect opium artifacts in the early 2000s and ultimately became addicted, described the process in detail, which is a method still used in many places today. He begins by explaining that smokers lie on their sides because that's the best way to smoke it. The reasons for the complex technique is that opium releases its alkaloids at a lower temperature than tobacco.

As mentioned before, it isn't smoked, it's *vaped*. Before opium is ready for "smoking" the latex has already been subjected to an extremely complex and painstaking process to transform it into a thick molasses-like syrup.

An opium pipe resembles a simple, roughly eighteen-inch-long bamboo flute with a covered bowl that has a tiny hole on the under-side. It's fixed to the pipe with an airtight fitting. The smoker (or his attendant) puts a little blob of that syrupy stuff on a steel needle and holds it above the chimney of an oil lamp. This makes it bubble and turn golden brown. Next, he or she rolls that mass into a small, pea-size ball. He then smears it over the top of the bowl, "cooking" it again, and uses the needle to push some down into that little hole. (It's the position of the smoker, pipe, and oil lamp that makes it possible for a skilled smoker to control the heat most efficiently.) Depending on whether the smoker is a "long draw" or "short draw" style user, he can inhale the whole amount in one toke or several—heating and pushing more opium into the hole as he inhales. (Anyone who has become fixated on brewing the "perfect" cup of coffee, rolling a "perfect" cigarette, or making the "perfect" cocktail, can at least have a sense of how the process could become an integral part of the habit.)

The entire process can be even more elaborate but it's how it was done then and is still done today. Regardless, it couldn't be managed by someone totally "out of it."[17]

## MYTH #7:

### *CONFESSIONS OF AN ENGLISH OPIUM-EATER* WAS WRITTEN BY A DRUG-CRAZED BRITISH WRITER WHO TRAGICALLY DIED YOUNG FROM HIS ADDICTION.

British essayist Thomas De Quincey, whose memoir is considered a seminal work of nineteenth-century literature, lived a long, albeit troubled, life.

De Quincey's lifelong addiction often gave him pause—and, at times, caused him intense physical and emotional suffering. Although it would be an overstatement to say he regretted it, his is without a doubt a cautionary tale. In particular, it raises a question that all drug- and alcohol-using creatives ask themselves: Does it help or hinder their creativity? And, if it does help, is the damage to health worth the possible reward?

Born in 1785, young De Quincey proved his dedication to rebelliousness and nonconformity at an early age. He ran away from school at seventeen and ended up in London where, refusing to take an allowance from an uncle, he did little but read and live hand-to-mouth

---

17   Martin, *Opium Fiend*, 66.

with a sixteen-year-old prostitute, with whom he allegedly had a platonic relationship. A year later, tired of a life of poverty, he agreed to be supported by his family and completed college and law school, although he refused to take the exams needed to get his diplomas.

While the reference to the Turkish "eating" habit seems to have worked better in his title, De Quincey actually drank his opium. He bought his first bottle of laudanum tincture when he was twenty to treat various aches and pains including a bad toothache.[18] It was love at first drop: "Here was a panacea...for all human woes," began the experiences he describes in *Confessions*. "Here was the secret of happiness, about which philosophers had disputed for so many ages, at once discovered; happiness might now be bought for a penny, and carried in the waistcoat-pocket; portable ecstasies might be corked up in a pint-bottle."[19]

At first, De Quincey only took the drug every few weeks, when he wanted a break from his obsessive writing. But by the time he was thirty years old, he was doing laudanum daily—so much he "could have bathed and swum in it," so much that he admitted he was at times virtually incapable of writing or even reading. But his dreams, which he eventually related in cinematic detail, are a testament both to his innate writing genius and opium's ability to unleash heavenly as well as unearthly sights and sounds.

De Quincey spent much of his professional life hiding from creditors; getting into petty literary spats with his editors; maintaining a love-hate relationship with his friend, mentor, and drug-literature rival Samuel Taylor Coleridge; and dealing with all sorts of physical aches and pains.

---

18  Self-reported descriptions of why people take painkillers are always suspect. De Quincey's "excuse" was the same as Sertürner's.
19  De Quincey, *Confessions*, 67.

He was also afflicted with a lethal combination of spiritual, creative, and British hubris that enabled him to state unequivocally that opium's wisdom wasn't accessible to everybody and, in any event, its visionary power was wasted on the common man. "If a man whose talk is of oxen should become an Opium Eater," he wrote, "the probability is that (if he is not too dull to dream at all) he will dream about oxen." Later, he speculated "whether any Turk, of all that ever entered the paradise of opium-eaters, can have had half the pleasure I had. But, indeed, I honour the barbarians too much by supposing them capable of any pleasures approaching to the intellectual ones of an Englishman."[20]

De Quincey appeared to be blissfully unaware of the literary price he paid for his breathtaking narcissism: i.e., when his drug-fueled visions are totally over the top, the writing itself hits rock bottom— combining the most self-indulgent features of an addict's convoluted thinking, circular logic, and insistent denial.

When trying to quit, De Quincey "lies under a world's weight of incubus and nightmare," while making "unexampled efforts of self-conquest," and walking "on a solitary path of bad repute, leading wither no man's experience could tell me." In spite of all his alleged superhuman suffering through withdrawal, he managed to achieve just six months or so of total abstinence in his many decades of use— eventually priding himself on reducing his daily intake to a modest few hundred drops (down from a maximum of "eight, ten, or twelve thousand").[21] Even the part of his book entitled "The Pains of Opium" casts his experiences in such a romantic self-aggrandizing light that he makes his addiction appear transcendentally noble. And that's mild compared to his rhapsodic descriptions in the chapter entitled "The

20   De Quincey, 77.
21   Ibid., 110–111.

Pleasures of Opium." Perhaps it's a testament to his writing, if not his addiction, that he managed to inspire many of his contemporaries to try the drug even while describing its pains in excruciating detail. He was just one in a long list of writers in France, America, and England who were opium or morphine users in the 1800s.[22]

People have long disputed whether the insights that people experience when they do mind-altering drugs are illusory or, rather, reveal a deeper reality within the "illusion" of everyday life. De Quincey clearly believed the latter—that the dreams the drugs release come from far deeper and more profound levels of reality than most ordinary mortals ever experience.

In modern terms, his early opium dreams have the markings of lucid dreaming (or, as he suggests, the unquestioned daydreams of a child).[23] His later opium dreams—for which he is rightly most famous—read like extraordinarily rich past-life regressions or near-death experiences.

It's here that his writing, while still afflicted with the excesses of 1800s Romanticism, soars as if he were liberated from his earthly arrogance long enough to let his natural talent and humanity emerge. Describing a vision he refers to as the "tyranny of the human face," he writes, "The sea appeared paved with innumerable faces, upturned to the heavens; faces, imploring, wrathful, despairing; faces that surged upwards by thousands, by myriads, by generations: infinite was my agitation; my mind tossed, as it seemed, upon the billowy ocean, and weltered upon the weltering waves."

---

22   Including: Louisa May Alcott, Charles Baudelaire, Aubrey Beardsley, Elizabeth Barret Browning, Robert Browning, Samuel Coleridge, John Keats, Rudyard Kipling, Edgar Allan Poe, Sir Walter Scott, Thomas Shadwell, and Oscar Wilde.
23   He poignantly compares his experiences to those of a child who once told him, "I can tell them to go and they go; but sometimes come when I don't tell them to come."

At one point, De Quincey imagines himself writing for people twenty, thirty, or even fifty years in the future, to "the many (a number that is sure to be continually growing) who will take an inextinguishable interest in the mysterious powers of opium. For opium *is* mysterious; mysterious to the extent at times, of apparent self-contradiction."

Thomas De Quincey took opium for fifty years, dying at age seventy-four.

# PART VI

Laws and Disorder

# Chapter 25

## America's First Failed Drug Laws

And so, we leave the rampant greed, unjustified wars, misunderstood scientific breakthroughs, and often brilliant if overwrought literature of the 1800s and move on to the rampant greed, unjustified wars, misunderstood scientific breakthroughs, and brilliant if overwrought literature of the 1900s.

Remarkably, there was not a single drug law on the books in America until 1875, when San Francisco's Board of Supervisors passed an "Opium Den Ordinance" that made it a misdemeanor to operate or visit an opium den. In the almost 150 years since then, local, state, and federal legislators have tried every form of penalty at their disposal—tariffs, taxes, fines, and imprisonment, even mandatory minimum sentences—to respond to the spread of opioids. In the process, while America may not have become a police state, it has frequently resembled a Keystone Cops state: no matter how hard the government chases after the problem, it never goes away. This is not to belittle the efforts of lawmakers and law enforcement officials. Rather it's a reflection of the ultimate inability of laws to provide a long-term solution to the ever-changing patterns of drug use, addiction, and overdose.

The 1875 law set the standard for several decades of California laws against *smoking opium*,[1] based on inaccurate or partial assumptions: that only Chinese-Americans would be interested in importing opium that had been processed to be smoked, that opium was only smoked in Chinese opium dens, that any Americans who smoked opium only did so at these opium dens and only because the Chinese had corrupted them, that Americans wouldn't consider processing their own opium for smoking, that American opium dealers wouldn't do business with the Chinese, and that importers and sellers wouldn't be able to easily find a way to get around the new laws.[2]

Each of these assumptions proved to be wrong.

In other words, the San Francisco law reflected the same kind of ethnic prejudice that many drug laws and enforcement strategies suffer from today. By targeting certain populations disproportionately, these laws reinforce stereotypes and discrimination, while doing little to solve the problem.

Just as Americans didn't really mobilize against the current opioid epidemic until it began to impact communities of every size and every socio-economic group in the country, San Francisco's city fathers weren't spurred into action until they became aware of the fact that "within three blocks of the City Hall [stood] eight opium smoking

---

1   The phrase *smoking opium* refers to crude opium prepared to be smoked by a traditional method of boiling and filtering out coarse impurities that sink to the bottom, and then simmering until it's a thick paste that is sun dried in molds until it hardens.

2   The *Los Angeles Daily Herald* (Nov. 19, 1875) referred to opium dens as "Mongolian death pens." Although, interestingly, in the ongoing argument of which was worse—alcohol or opium—the Los Angeles paper came down firmly on the side of the former: "There is another class of dens found all over San Francisco that are a thousand times more damaging to public morals, more destructive to health and productive of more misery than those frequented by the Chinese. They are the back alley and cellar dens where laboring men spend their small means in the purchase of the most villainous of alcoholic drinks that ever maddened the brain."

establishments, kept by Chinese, for the exclusive use of white men and women." These places were not just patronized "by the vicious and depraved," but they were "nightly resorted to by young men and women of respectable parentage."[3]

Few people were concerned about this blatant attack on the civil liberties of Chinese immigrants, although eventually the California Supreme Court, in ruling on an ordinance in Stockton, acknowledged that "the object of the police power is to protect rights from the assaults of others, not to banish sin from the world or make men moral."[4]

In 1881, California passed a *statewide* law that made operating an opium den a misdemeanor, but it was only honored in the breach. The locations of opium dens not only remained common knowledge, but actors pretending to be dope-crazed Chinese were hired by tour companies to give their customers a thrill—a taste of the "real" San Francisco.[5] Although the tourists' shock and awe was only one tip of the hysterical iceberg that eventually misinformed opium laws all over the country, it helped further inflate the image of tawdry opium dens as the tourists returned home with shocking tales about life in the belly of the opium beast.

San Francisco's elders, realizing that even the statewide opium den crackdown wasn't working, decided to charge wholesale opium dealers a license fee to unload the product in San Francisco harbor. So, while it was perfectly legal for an American to take possession of the drug, it was a misdemeanor for a Chinese-American to run a place where opium was smoked. This would be like making liquor legal but outlawing bars, or, worse, charging drug kingpins a fee and limiting arrests to street dealers.

---

3   Cited in "125 Years of the War on Drugs."
4   Ibid.
5   One of the tourists (and, one could argue, propagandists) was Mark Twain, who said that "the stewing and frying of the drug and the gurgling of the juices in the stem would well-nigh turn the stomach of a statue." See Krskag, "The Form of Money."

Eventually, the state tried to rescue San Francisco and itself from the unintended consequences of its lukewarm attempts to restrict the opium trade by simply outlawing all nonprescription opium and cocaine sales in 1907—at which point California authorities started arresting unsuspecting users, even some who weren't Chinese. The city quickly got into the spirit of the new statewide prohibition by staging huge bonfires, reminiscent of Commissioner Lin's forays into mass opium destruction decades before. One, which took place in front of San Francisco's city hall, was fueled by confiscated drugs and 500 opium pipes.[6] Fortunately for posterity's sake, Mayor P. H. McCarthy recognized a priceless antique when he saw one and had a few set aside to be displayed at the Golden Gate Museum.[7]

While a few cities in California and around the country tried various strategies to crack down on opium abuse, the federal government's only attempts to regulate the trade in the nineteenth century were through tariffs. In 1880, America and China signed the Angell Treaty which prohibited all Chinese residents from *importing* the smoking variety of opium from China. This wasn't just for America's benefit. The Chinese government was equally serious about cracking down on the opium trade in *their* country.

Americans, however, remained convinced that *smoking opium* was the source of all evil and that the Chinese government was behind it and that Chinese were the only ones importing it. So, the treaty neglected to make it illegal for *Americans* to import smoking opium—which they had been doing all along in ports such as New York, New Orleans, Los Angeles, and the northwest, as well as at various points along the Mexican border. So, much like a teenager getting an adult

---

6 *Colliers,* "A San Francisco Public Bonfire."
7 "Opium Museum."

to buy him or her liquor or cigarettes, all a Chinese dealer in America had to do was find a white American co-conspirator who would import opium for them in exchange for a healthy markup.

However, even though Americans could still import smoking opium, as part of the Angell Treaty, Congress levied high import tariffs on it. Fortunately for the dealers, the government staggered the implementation of these tariffs over seven years so dealers had plenty of time to stock up before they went into full effect. During the 1870s, imports of smoking opium had typically ranged from about 40,000 to 75,000 pounds a year. In 1882, however, importers brought in 106,000 pounds. In 1883, imports grew again to a staggering 298,000 pounds. As the law began to be implemented in 1884, imports did drop dramatically, and it seemed the tariffs might work. But the coming years revealed that imports had dipped simply because suppliers had stocked up in advance. By 1886, imports had returned to almost 50,000 pounds per year.[8]

Over the next decade, Congress repeatedly tried to fine-tune the tariffs—raising them from $6 per pound slowly to $12 per pound, only to see their actual revenues decrease as the higher taxes simply led to more smuggling. They finally settled back on $6 per pound as the "sweet spot," at which it was more cost-effective for dealers to import the opium and pay the tax than smuggle it in. By the turn of the century, imports were again back to almost 130,000 pounds a year.

The next significant anti-opium effort in America focused on the patent-remedy business. The 1906 Pure Food and Drug Act mandated that food and drugs that crossed state lines had to be correctly labeled. In addition to letting them know that there was an excessive amount

---

8  Redford and Powell, "Dynamics of Intervention in the War on Drugs." Drug import figures are always unreliable (due to smuggling), but the trend of the numbers speaks to the futility of those laws and regulations.

of mouse droppings in their favorite foods, consumers were alerted to the fact that their magic elixirs included opium and other narcotics, a revelation that did not necessarily discourage them, since in many cases, they knew it already and it only served to assure them they were getting the potency of painkiller they wanted.

By 1909, opium was increasingly being seen as a national crisis. However, it took international pressure to force Congress to pass a national "Smoking Opium Exclusion Act."[9] Congress had no intention of sending federal agents to break down the doors of law-abiding white citizens, storm into their bathrooms, and smash every patent remedy in sight. That, and the fact that the "exclusion" still only applied to imports of *smoking opium*, left the barn door so wide open that dealers could stroll in and out as they pleased. In addition, the law did nothing to prevent an American from importing *crude* opium, processing it into smoking opium, and selling it to Chinese dealers as they had been doing since the Angell Treaty.

American importers and processors also sold it themselves directly to the increasing number of Americans who were smoking opium in the safety and privacy of their own homes—if they hadn't already switched over to the stronger, cleaner morphine or new-fangled heroin, neither of which had been regulated yet.

The law's weaknesses were no secret, and between the Shanghai Conference in 1909 (the first significant international effort to deal with the opium problem) and the approaching Hague Conference in 1912, Congress was pressured by anti-opium crusaders to appear to get serious about helping China in its centuries-long struggle to keep opium from

---

9    The major objection to the bill came from proponents of states' rights who felt states should be able to continue unsuccessfully regulating the behavior of their own citizens.

beating Confucian and Taoist ideals of moderation into submission. So in 1914 they passed the Harrison Act. While considered the grandfather of American drug legislation, the Harrison Act didn't actually prohibit opium. Its key stipulation was that anyone who dispensed narcotics in any possible way or used them in research or education had to register with the government and pay an annual "occupational tax" of $1 to $24. It also made it illegal to possess a drug that hadn't been "issued for legitimate medical uses by a physician" and closed a loophole that made it possible for smugglers to avoid paying the 1909 taxes by simply transferring tax stamps (proof of tax payment) from container to container.[10] But, like many so called "sin taxes," perhaps the most dramatic effect of taxing opium was to change the cost-benefit equation for organized crime; the income/social-benefit equation for society; and the fear-craving benefit for the addict.

The act also earned the government a modest amount of income. By 1916, almost a quarter million doctors, pharmacists, dentists, wholesalers, importers, researchers, and educators had registered, as had some veterinarians, although it's unclear whether any of those health professionals or their four-legged patients were ever prosecuted to the full extent of the law.

However, the act's most critical, confounding, and controversial clause was that, while it allowed doctors to prescribe opioids to treat certain illnesses, relieve pain, and even wean an addict, it made providing *maintenance* doses a federal crime. At the time, there were about 250,000 opium addicts, many of whom had become addicted

---

10   While the new law essentially was retroactive (any opium shipments inspected after the law was implemented was presumed to have been imported after the law was passed) it didn't take a master criminal to figure out how to remove the tax stamps from any empty containers and put them on existing stock before the feds showed up.

while being legally prescribed the drug and were best able to live productively only as long as they continued receiving these maintenance doses. In other words, prescribing a drug to treat addiction as a chronic disease was now illegal because addiction was not seen as a disease, a troubling state of affairs that continues today.[11]

In the end, the prohibition on maintenance treatment led to the arrest of about 25,000 doctors and, while only a few went to jail, many lost their reputations and life's savings.[12] In addition, as a result, those whose addictions were incurable or hadn't yet entered treatment before the law took effect resorted to the black market, where they paid higher prices for drugs of unreliable purity and potency.

In 1922, Congress tried again to restrict the flow of drugs with the passage of the Jones-Miller Act, a.k.a., the Narcotic Drugs Import and Export Act, which made it illegal to import any narcotic that wasn't required for medicine or science and, additionally, put the burden of proof on the defendant to show he or she legally possessed a drug. In what proved to be the first salvo in the battle to establish mandatory sentencing guidelines, a first offense meant five to twenty years in jail and a $20,000 fine. Subsequent offenses were good for ten to forty years and another $20,000 out-of-pocket. Most significantly, anyone over eighteen who knowingly provided heroin that had been unlawfully imported to anyone under eighteen had to serve from ten years to life in prison, and could even be given the death penalty if the jury deemed it appropriate.[13]

---

11  Libby, *The DEA's War on Doctors.*
12  Hogshire, *Opium for the Masses*, 226. One of the most interesting unintended consequences of the act was that the newly formed American Medical Association lobbied in support of it, which helped solidify them as the medical establishment's "voice of authority" during future drug-related congressional negotiations.
13  See Cantor, "The Criminal Law and the Narcotics Problem," and Libby, *The DEA's War on Doctors.*

# Chapter 26

# Drug Hysteria and Race-Based Enforcement: The Harry Anslinger Story

On December 17, 1914, decades before President Dwight David Eisenhower famously warned the country of the danger of the "military industrial complex," President Woodrow Wilson signed the Harrison Act and laid the groundwork for a "narcotics industrial complex" that now costs the United States approximately $64 billion a year—a total of more than $1 trillion since President Richard Nixon declared the "War on Drugs."

Since the Harrison Act didn't make drugs illegal, it simply taxed them. The first department involved in what has become today's maze of drug enforcement was the Bureau of Internal Revenue (now the IRS). Commissioner William H. Osborn was authorized to appoint and employ as many "agents, collectors, inspectors, chemists, assistant chemists, clerks, and messengers" as he needed—and given a grand total of $150,000 to do the job.[1]

Collecting taxes on drugs, however, was the least of Osborn's problems. He had also just been put in charge of the new income tax, which was called, even back then, "the largest, and most intricate

---

1   Harrison Act, Pub. L. No. H.R. 6282 (1914).

revenue measure ever conceived in the history of the world."[2] So, he relegated oversight for ensuring compliance with the new drug law to his Miscellaneous Division, which also oversaw taxes on margarine, adulterated butter, flour, playing cards, and cotton futures.

Osborn's 162 narcotic agents immediately began arresting doctors and pharmacists who didn't have official tax stamps. Four months into the enforcement campaign, however, he had received enough reports from the field to understand just how punitive the law was for existing addicts, so he urged Congress to amend it to take into consideration "the sufferings of those unfortunate citizens addicted to the use of the narcotics... who, in a great many instances, are financially unable to obtain necessary treatment at hospitals or sanitariums, and in other cases because of advanced age or physical infirmities, cannot be deprived of the drugs without endangering their lives."[3]

His plea would fall on deaf ears. This left Osborn with an impossible task, a shrinking budget, and a government that badly needed money to prepare for the world war into which they were inexorably being drawn. In a desperate move, the bureau began to give its collectors— who were paid a pittance—a bounty on the amount they collected in taxes, up to $4,000 for collections of a million dollars (4/10 of 1 percent). While $4,000 was a significant incentive, in some cases, agents found it even more profitable to make "arrangements" with anyone they discovered trying to evade the tax.

When Prohibition became law in 1919, the Bureau of Internal Revenue assigned narcotics enforcement to the new Prohibition Unit. The same year the Supreme Court upheld the Harrison Act, which had been challenged on humanitarian grounds, and ruled that prescribing

---

2   "Urges Taxpayers to Aid Government," *New York Times*, June 10, 1917.
3   Commissioner of Internal Revenue, "Annual Report 1915."

"maintenance doses" to addicts simply wasn't acceptable medical treatment. Based upon that ruling, agents now started raiding the hundreds of addiction clinics run by doctors who were willing to provide maintenance doses to patients unable to successfully complete withdrawal.

Some cities and towns—including New Orleans, Atlanta, Los Angeles, Cleveland, Memphis, and Houston—fought back by establishing community addiction clinics, hoping that if private clinics were illegal, the federal government might permit public ones. The federal government soon disabused them of that notion and began shutting the public clinics down. All but one was closed by 1921 and the last one, in Shreveport, Louisiana, closed its doors on February 10, 1923.[4]

Narcotic prohibition was overshadowed by alcohol prohibition during the 1920s, and the difference between the laws covering them was significant. Under prohibition, it wasn't legal to manufacture or sell alcohol but drinking itself was not regulated. In the case of nonprescribed opioids, on the other hand, they were illegal—no loopholes and no exceptions. By the late 1920s, about a third of prisoners in federal penitentiaries were there for violations of drug laws, far more than the percentage for liquor violations. Regardless, the use of opium and rate of addiction continued to rise and the government eventually opened special narcotic prisons in Fort Worth, Texas; Lexington, Kentucky; and McNeil Island in Washington State. In addition to housing those who had actually broken laws, the one in Kentucky also welcomed users who wanted to voluntarily commit themselves to get clean.[5] The farm work was, allegedly, rehabilitative, the food was decent, and the entertainment was top-rate since residents included

---

4   Courtwright, "A Century of American Narcotic Policy."
5   Musto, *The American Disease*, Ch. 8.

famous jazz musicians like Sonny Rollins, Chet Baker, and Elvin Jones. As trumpeter Dizzy Gillespie put it, "Cats were always getting busted with drugs by the police, and they had a saying, 'To get the best band, go to [Lexington] Kentucky.'"[6] Unfortunately, recidivism rates were enormous. For instance, up to 90 percent of those released from the Lexington "drug jail" relapsed.[7]

\* \* \*

While alcohol prohibition was controversial, prohibiting the use of opium was widely popular.

Newspaperman William Randolph Hearst—a self-proclaimed moral authority and master of public opinion—only believed in drug prohibition, and he leveraged the most famous proselytizers and propagandists of the time for his cause, in particular Captain Richmond Hobson, whose fame was based on rather questionable heroic acts in the Spanish-American War and even more questionable claims about the dangers of opioids. Best described as a man of "virtually unlimited moral indignation,"[8] Hobson began his career railing against the dangers of the demon rum, until that "crusade" dried up thanks to Prohibition. Seamlessly transitioning from one work of the Devil to another, he spent the rest of his life feeding the country's innate loathing of narcotics with apocalyptic visions of civilization's imminent demise.

Hobson spearheaded the International Narcotic Association in 1923, the World Conference on Narcotic Education in 1926, and the

---

6  Quoted in Filan, *Power of the Poppy*, 171.
7  Courtwright, "A Century of American Narcotic Policy."
8  Quoted in Courtwright, *Dark Paradise*, 32.

World Narcotic Defense Association in 1927. On March 1, 1928, he celebrated the closing of the new annual Narcotic Education Week, in a speech over NBC radio that began with some solid history and science that it proceeded to smother with over-the-top rhetoric.

After acknowledging the importance of education and treatment—and even the fact that the drug problem started with "white men" forcing opium onto the "unwilling peoples of Asia"—he described withdrawal in brazenly graphic terms: "A condition of torture sets in. The muscles seem to become knotty. Cramps ensue in the abdomen and viscera, attended frequently by vomiting and involuntary discharge of the bowels. Pains often succeed each other as though a sword were being thrust through the body...the degeneration of the upper brain is so swift that the elements of character crumble in a few months." To make sure the public fully understood the threat drugs posed he then abandoned any semblance of scientific facts and went into full "fire and brimstone" mode, claiming that addiction is "more communicable and less curable than leprosy" and that, "drug addicts are the principal carriers of vice disease, and with their lowered resistance are incubators and carriers of the streptococcus, pneumococcus, the germs of flu, of tuberculosis and other diseases." Insisting that heroin addicts would lie, steal, rob, and even commit murder to get their fix he warned his listeners that they were "now in the midst of a life and death struggle with the deadliest foe that has ever menaced [their] future. Upon the issue hangs the perpetuation of civilization, the destiny of the world and the future of the human race."[9]

\* \* \*

---

9    "The Struggle of Mankind Against Its Deadliest Foe." Also see Musto, *The American Disease*, Ch. 8.

After Prohibition ended, drug enforcers finally had a new agency they could call their own, the Federal Bureau of Narcotics. This launched the career of its first commissioner, Harry Anslinger, the person most synonymous with the phrase "war on drugs"—in fact, the first person to use it—and likely the first person, outside of any royal family, to be referred to as a "czar."[10]

Technically, Anslinger's powers were limited to enforcing the Harrison Act, Jones-Miller Act, and a 1924 amendment to Jones-Miller that banned the production and sale of heroin. But the broad language of those laws could be stretched to justify arresting anyone involved in any way with the importation, production, sales, or use of opioids, unless legally prescribed by a registered physician. Between 1930 and 1962 Anslinger established the standards that continue to serve as basic tools of the trade for America's drug enforcement, such as dramatic drug busts, harsh penalties, and questionable data.

Anslinger traced his hatred for drugs to two events in his childhood. One was the day he heard the scream of a neighbor's wife in withdrawal, screams that only ended when he returned from the local pharmacy with the medicine her husband had sent him rushing off to get—the kind of medicine even a young boy could buy legally in those days. At the time, he may not have known he was "trafficking" in morphine, but he claimed he had never forgotten those screams.

While that experience instilled a horror of drugs, his other seminal moment gave him an inkling of the true root of the problem. As a teenager, he worked alongside Italian immigrants, maintaining the railroad tracks near his hometown of Altoona, Pennsylvania, and one day he saw the badly beaten body of a co-worker named Giovanni in a ditch.

---

10    He actually used the word "war*fare*." See "'One Thing You Can Say for the War on Drugs...Is We Gave It a Fair Shot.'"

Somehow the man survived and nervously confessed that he had been attacked by "Big Mouth Sam," who had been trying to extort money from him. Harry confronted Big Mouth Sam and threatened to kill him if he ever attacked Giovanni or any other railroad worker again. Later, he would claim that's when he learned that the "Black Hand" (a.k.a., Mafia) had begun infiltrating American society.[11]

Anslinger used these stories of traumatic drug-related childhood events to persuade people he had unique insights into the drug trade that legitimized the sense of unquestioned self-righteousness with which he ran the Federal Bureau of Narcotics. However, these and other details of his early life were based on Anslinger's own reminiscences—which featured the kind of exaggerations that would become his stock-in-trade. Anslinger was not even above embellishing his own résumé when it suited his purposes—once, for example, claiming falsely that he had earned a degree from the University of Maryland Law School.

There was, indeed, something Jekyll-and-Hyde about Anslinger's personality. He's been called everything from a wily, efficient drug hardliner to an unscrupulous fear-mongering propagandist to a "social entrepreneur" who would do whatever it took to keep his bureau in business and grow its "brand."[12]

There remains serious disagreement in scholarly as well as political circles about how successful Anslinger really was in reducing drug sales and use in America. Some consider him the unsung hero of twentieth-century drug enforcement: "For 40 years, very little happened in the world of drugs or international affairs that Anslinger did not know about," one biographer boasted. "At times it

---

11   See McWilliams, *The Protectors*, 195.
12   Hari, *Chasing the Scream*, 305.

appeared that he was virtually omnipresent…chasing *Mafiosi* across Europe, condemning communist aggression in testimony at congressional hearings, or coordinating a nationwide 'bust' of pushers and peddlers."[13]

Others claim almost the exact opposite: "Anslinger concentrated on apprehending junkies and small-time pushers but leaving the top of the trade untroubled. He never indicted or even investigated organized crime. In return, the bosses would stage spectacular pseudo-hauls from time to time which were widely trumpeted in the media, Anslinger being photographed standing in a Napoleonic pose next to crates labelled 'heroin.' Such crates never existed: even the dimmest drug boss was not that dim."[14] To skeptics like these, Anslinger was simply a figurehead who "survived by bluster and obfuscation."[15] In other words, organized crime bosses were quite content to help him persuade others, and perhaps even himself, that he was making significant progress in his war on drugs as long as the majority of their product was able to stay under the radar.

Anslinger understood that fighting the drug trade involved international cooperation, so he lobbied enthusiastically for the United States to join INTERPOL.[16] When that didn't happen, he continued to attend any meeting he thought might lead to international restrictions, and, in 1933, played a major role in creating an alliance of twenty-five nations that ratified a treaty in which they pledged to

---

13  Owen, "Standing in the Shadows: The Legacy of Harry J. Anslinger."
14  Dormandy, *Opium*, 250.
15  Davenport-Hines, *The Pursuit of Oblivion*, 362.
16  The United States did join for a little while after World War II. However, by then, J. Edgar Hoover was spooked by the presumed presence of communists in every foreign government (as well as his own), and pulled out. It's not clear whether Hoover and Anslinger respected or suspected each other (undoubtedly a little of both) but in terms of spymastery Hoover would always get the final word.

import only enough narcotics for medical use, thereby making less available for diversion.[17]

His most famous international success—or at least the one he took the most credit for—was exposing the legendary "French Connection."[18] In the late 1940s, his agents discovered that large quantities of opium were finding their way from Lebanon and Turkey to Corsican gangs who smuggled it into Marseilles, where it was processed into heroin and morphine and shipped to America.

Uncovering the heroin's source and roundabout route to America should have made ending the operation straightforward. Unfortunately for Anslinger, the CIA was, at the same time, supporting the drug-dealing Corsicans to get their help keeping the port of Marseilles from being taken over by communists in France. Thanks to complications like these and the general resilience of international drug rings, variations of the French Connection outlived Anslinger—at least until the early 1970s when the Turkish government began restricting poppy cultivation and the major source of supply shifted to the Far East's Golden Triangle.

Ultimately, it wasn't as important to Anslinger whether his campaign against drugs needed to be fought overseas or at home, as long as his Federal Bureau of Narcotics had a place to wage its war and could claim it was winning. As he once said, "Every victory leads to a new field of battle."[19] In other words, regardless how many successes he could boast of, there would always be the need for his agency to grow in power, stature, and budget.

In terms of his own agenda, some of Anslinger's greatest successes

---

17   Courtwright, "A Century of American Narcotic Policy."
18   McWilliams, "Unsung Partner against Crime," 225.
19   Drug Enforcement Administration, "DEA History—The Early Years."

were back in America. While he may not have done much to reduce drug use over the long term, he achieved several significant legislative victories, including the Uniform State Narcotic Drug Act, which fostered collaboration between federal agents and police in different states (each of which had its own specific laws). The act set a standard for state laws that forbade anyone to "manufacture, possess, have under his control, sell, prescribe, administer, dispense, or compound any narcotic drug except as authorized."[20] It specifically mentioned cocaine, opium, and its derivatives, as well as synthetically produced opioids. Cannabis was only mentioned in the fine print of subsection fourteen. Eventually adopted by twenty-seven states, the act may have simplified interstate enforcement but it also made it easier for dealers who no longer had to look up a whole new set of laws to do their risk-profit calculations every time they decided to expand their empires.

Anslinger's next major legislative victory was the passage in 1951 of the Boggs Act, which imposed mandatory minimum sentences for possession, decades before they became popularized in the 1990s as the panacea for persistent crime.

He also facilitated the passage of a third law, the Narcotic Control Act in 1956, which increased those penalties and said a jury had the discretion to ask for the death penalty when someone over eighteen was convicted of selling heroin to someone under eighteen.[21] Perhaps the most malicious (and constitutionally questionable) of Anslinger's laws, the Control Act, made it legal to arrest people without a warrant, as long as there was a "reasonable" suspicion that someone was breaking a narcotics law. In some situations, it was even possible to

---

20   "Addiction A–Z."
21   Sacco, "Drug Enforcement in the US," 4.

prosecute people for breaking narcotics laws without any actual proof of purchase or possession—helping to earn its reputation as "one of the most oppressive pieces of legislation ever passed by Congress."[22]

\* \* \*

As difficult as passing drug laws is, enforcing them effectively, consistently, and fairly has proven to be virtually impossible. The 1875 San Francisco ordinance that focused on Chinese users was simply the first example of legislation in which race, ethnicity, and culture would not only prove to be as damning as sales or possession, but would make a huge impact on public opinion of both the ethnicities and inebriants involved.

Anslinger took this type of racial profiling to a new level. By putting the blame for drug problems on other countries, he made a major contribution to the American tradition of conflating its drug wars with its cold wars—even when, as with the French Connection, they weren't necessarily aligned.

In the early 1950s, for example, he claimed that communist China was producing more than 4,000 tons of opium a year and using it to flood the market with heroin in order to fund its campaign of world domination and cause moral decay throughout the free world. That was a particularly ironic accusation considering the West's treatment of China during the Opium Wars—and shamelessly duplicitous in light of communist China's attitude to opium. In 1949, as many as 20 million Chinese, or more, were addicted to opioids and Chairman Mao Zedong's communist government responded by banning the

---

22  Dormandy, *Opium*, 251.

drug severely and completely,[23] razing opium fields, and paying fiery homage to Commissioner Lin by burning a ton of opium in Canton. China also placed addicts on strict detox schedules—and threatened severe punishments for those who didn't comply. In addition, almost 1,000 big-time traffickers were executed. Anslinger's claim about Chinese complicity in the American drug trade[24] was such a bald-faced lie that, in the 1960s, his own former agency issued a report that China hadn't been exporting any opioids at all.[25]

Indeed, throughout his career, Anslinger would change his metrics faster than a Ponzi schemer trying to cover his tracks, in order to convince the president and Congress that he was making progress and that they should increase his budget accordingly.[26] His creativity, however, wasn't limited to numbers, or even opioids. Early on, he recognized the potential of marijuana to shape public opinion by calling it a "gateway" drug. Back in the 1930s, he had made a major contribution to the hysteria reflected in films like *Reefer Madness* and *Assassin of Youth*. Ironically, those very films proved to be "gateway drugs" of sorts in the 1960s. It only took a new user a couple of tokes to see through the patently fearmongering nature of the films.

No wonder Anslinger once said, "The only persons who frighten me are the hippies."[27]

Anslinger undoubtedly would like to be remembered for his self-proclaimed legends from childhood (rescuing the screaming farmer's

---

23   Estimates vary widely. Griner, "China's New Opium Wars," says 20 million or 5 percent in 1949, whereas Booth, *Opium*, says 10 percent of the population (40 million) at the end of the 1930s, and that the numbers didn't go down during the 1940s.
24   Griner, "China's New Opium Wars."
25   Marshall, "Cooking the Books."
26   Courtwright, *Dark Paradise*, 118ff.
27   McWilliams, *The Protectors*, 187.

wife and the badly beaten Italian railroad worker). Unfortunately for his legacy, his most striking claim to fame, or, perhaps infamy, was his blatant cultural bias, if not outright racism—a character trait that, of course, still plagues modern drug enforcement. He unapologetically divided the world into us and them, good and bad, right and wrong—and always black and white. These days, in the eyes of some of those who see the world Anslinger's way, Muslims may have "replaced" communists as the enemy within; Afghanistan and Mexico may have joined China as the countries we blame for *our* addictions; and, in terms of drugs, Hispanics may now be feared as irrationally as African-Americans.

Tying his hardline approach to drugs to the country's hardline approach to communism gave Anslinger a troublesome ally in the early 1950s—Joseph McCarthy, the junior senator from Wisconsin, a man whose disregard for the truth destroyed many peoples' reputations and threatened Anslinger's power.

After earning a Distinguished Flying Cross in World War II by lying about his number of aerial missions, forging a letter from his commanding officer so that he would get a letter of commendation, and lying about his opponent's age during his first election, Senator McCarthy became most famous for waving a piece of paper in the air that he claimed had the names of 205 Americans who belonged to the Communist Party. With some significant support from the FBI's J. Edgar Hoover, McCarthy destroyed the careers of writers, actors, labor organizers, scientists, politicians, and even some members of the military whom he or his allies on the House Un-American Activities Committee thought might threaten their power or perceived patriotism.

Until his fall from grace, thanks in large part to seven simple words uttered by a Boston lawyer named Joseph Welch—"Have you no sense of decency, sir?"—McCarthy's efforts to expose communists dovetailed nicely with Anslinger's efforts to blame China for the epidemic of heroin on the streets of America's cities.

Like many demagogues, however, McCarthy's behavior gave the lie to his beliefs because *he* was a drug addict. His addiction to morphine worried Anslinger, and rightly so, because anyone who found out would be in a position to blackmail McCarthy and undermine Anslinger's anti-Chinese rhetoric. So, he tried to persuade McCarthy to give up the drug. But after McCarthy told him no one was going to stop him, Anslinger arranged for the Capitol Hill Pharmacy to secretly provide his friend with high-quality morphine.

While Anslinger's thirty-year war on drugs undoubtedly saved the lives of some individuals, his racial prejudices tarnished his reputation in ways that, even allowing for 20/20 hindsight, can't be dismissed. The most blatant example was his disparate treatment of two of the nation's most famous celebrities in the 1950s: Judy Garland and Billie Holiday.

Anslinger didn't like stars. He knew they often ridiculed him because they felt their wealth and connections made them untouchable, and his distaste for them became something of an obsession.[28] Still, he treated Judy Garland with kid gloves. The star of *The Wizard of Oz* suffered from an eating disorder before the phrase was coined. Determined to make sure she lost her baby fat as she evolved from brick roads to adult roles, her studio, Metro-Goldwyn-Mayer, and Garland engaged in what she called a "constant struggle...whether or not to eat, how much to eat, what to eat. I remember this more vividly than anything else about my childhood."[29] In this, MGM had a trusted ally in the star's mom, Ethel, who had already been giving her daughter

---

28   Dormandy, *Opium*, 250.
29   Clark, *Get Happy*, 82.

pills to in the morning to wake up and at night to go to sleep. "Now," as one biographer put it, "Metro added diet pills, combinations of Benzedrine and phenobarbital, to that already potent mixture. The studio had found the ultimate weapon."[30] It was a weapon that worked not only to keep her weight off but also allowed her to stay "vivacious before the camera during the long shoot days."[31]

Soon she began to demonstrate the kind of unreliability—in terms of when she'd show up on set and what condition she'd be in—that Marilyn Monroe became famous for.

With informants planted all over Hollywood, Anslinger knew what drugs Garland was doing and where she was getting them, so one day he intervened by visiting the heads of MGM and insisting they send her to a sanatorium, saying, "I believed her to be a fine woman caught in a situation that could only destroy her."[32] He was told they had $14 million invested in her and had no intention of giving her the time off she needed. An unsuccessful suicide attempt—even if only a cry for help—finally persuaded them that the best way to protect their investment was to send her to rehab. Later, Anslinger would imply that he had played the major role in helping Garland get clean, but that may also have been just a story only he believed.[33]

---

30   Ibid, 83.
31   Brody, "Judy Garland's Hollywood Unravelling."
32   Clark, 277.
33   While Harry Anslinger might have liked the film that made her famous, he may not have been aware of the irony of Judy Garland waltzing through a field of poppies and then falling asleep with some strange creatures who had no brains, no heart, and no courage. There have been a wide range of interpretations of *The Wizard of Oz*'s symbolism over the years—a populist economic fantasy; a religious allegory; a feminist allegory (the author's mother-in-law was the suffragist Matilda Joslyn Gage); or a drug parable, which, in addition to the soporific poppies, has the following parallels: Oz = ounce; flying monkeys, etc. = bad acid trip; Glinda sprinkling Dorothy with snow to wake her up = cocaine; Lollipop = a gateway drug; and our personal favorite: that the Tin Man is a heroin addict who needs a fix [lubrication] to function. See Man, "Trippy Films."

Anslinger's reaction was exactly the opposite when he heard about a black woman singing about strange fruit hanging from poplar trees: "Black bodies swinging in the southern breeze." For him, Billie Holiday was an ideal symptom of the drug problem—and an irresistible target for his crusade. She was a woman. She was black. As a child, she scrubbed floors in a brothel. She was eleven when someone first tried to rape her. She made $5/client when she was pimped out at fourteen. She sang jazz—an improvised, undisciplined, free-form music with Caribbean and African influences that Anslinger referred to as sounding "like jungles in the dead of night." To make things even worse, she flaunted how much she drank and the drugs she took.[34]

The brilliant scholar and 1960s cultural icon Angela Davis makes the point that most common portraits of Billie Holiday "highlight drug addiction, alcoholism, feminine weakness, depression, lack of formal education, and other difficulties unrelated to her contribution as an artist."[35] Indeed, most creatives—writers, artists, musicians, actors, et cetera—are known first for their works. Their biography serves as background. With Billie—as with other celebrities who have dealt with addiction—it's often the other way around. Emphasizing Holiday's drug use enabled Anslinger to minimize her influence on the arts, social consciousness, and the growing civil rights movement—a way to keep the focus on her color not her creativity, on her addictions not her artistry.

When, by June 1959, Holiday was in New York's Metropolitan Hospital on her deathbed, weakened by liver and heart disease, Anslinger ordered agents from his Federal Bureau of Narcotics to storm into her room and arrest her for drug possession. They handcuffed her and put her under a police guard until a judge ordered the cuffs removed.

---

34   Hari, "The Hunting of Billie Holiday."
35   Davis, *Blues Legacies and Black Feminism*, 184.

Whenever she performed the song "Strange Fruit," she had the lights turned down. And when they came back on, she was gone. Billie Holiday died in that locked hospital room on July 17, 1959, at age forty-four.

Harry Anslinger died on November 14, 1975. He was eighty-three years old and suffered from an enlarged prostate and angina. At the end of his life, Anslinger took morphine for the pain.[36]

---

36   Hari, *Chasing the Scream*, 298.

# Chapter 27

# The War Nobody's Ever Won

On June 18, 1971, in a message to Congress, President Richard Nixon famously declared a war on drugs, calling them "Public Enemy Number One." Since then, eight administrations have signed drug laws and regulations that have, in the long run, proven as ineffective as those of the Chinese emperors centuries before.

Instead of heralding the end of America's drug problem, Nixon's pronouncement marked the beginning of fifty more years (and counting), during which narcotics would only get stronger and regulations less effective. The American prison system has been overwhelmed by drug users convicted of nonviolent associated crimes ever since. At the same time, the demographics of those doing opioids skew younger and penetrate deeper into America's heartland with each passing decade, and the number of overdoses are now at previously unthinkable heights.

During the decade before Nixon's speech, it seemed like the country was finally entering a new, more enlightened era of drug policy. Harry Anslinger had retired from the Federal Bureau of Narcotics—perhaps with a push from President Kennedy, whose advisers had succeeded in convincing him that addiction was a disease to be treated, not a crime to be punished.

Even though Kennedy was assassinated before he could see his ideas become law, President Johnson signed several laws that built on Kennedy's idea that, whenever possible, addiction, alcoholism, and other mental illnesses should be treated at community health centers rather than inpatient psychiatric hospitals, which, in many cases, were institutions of chaos and even abuse.[1] Shortly thereafter, Johnson's Katzenbach Commission recommended that the National Institute of Mental Health develop educational and informational materials as part of an overall harm-reduction campaign. It appeared that addiction was finally going to be treated as disease.

When President Richard Nixon was elected, however, things began to look less promising for treatment-based drug policies. Nixon had a reputation for being hard on crime, a reputation that his rhetoric indicated he was ready to live up to.[2] But, while declaring his war on drugs, he also showed that he understood the role of treatment.

"As long as there is a demand," he said in his speech announcing the war on drugs, "there will be those willing to take the risks of meeting the demand. So, we must also act to destroy the market for drugs, and this means the prevention of new addicts, and the rehabilitation of those who are addicted."[3]

---

1  One unintended consequence of this strategy was that it led to *insufficient* numbers of inpatient hospital beds for those who really needed them. This continues to be true due, in part, to Medicaid's limits on inpatient days for addiction and other mental-health diseases—limits that would never be tolerated for physical illnesses.

2  Except, perhaps, *for his own crimes*, particularly his involvement in the break-in of Democratic Party Headquarters (which, as with Russia's alleged interference in America's elections) pose as great a threat to national security as drug use. There's a disturbing pattern of administrations that claim to be hard on drugs, but undermine their own legitimacy by considering themselves above the law. In fact, Nixon once told interviewer David Frost, "I'm saying when the president does it, that means it's not illegal." See Kurtzman, "What Are the Most Ridiculous Richard Nixon Quotes?"

3  "Public Enemy Number One."

Nixon may be known as a law-and-order president, but as the quotation suggests, his legacy would prove to be more nuanced. While he signed major pieces of legislation that increased the size of federal drug-control agencies, he also eliminated virtually all mandatory minimum sentences left over from the Boggs Act and established the National Institute on Drug Abuse, a federally funded independent research institution that he tasked with developing scientific strategies to combat drug abuse and addiction. Under his Controlled Substances Act, the Department of Justice was authorized to "schedule" drugs on a scale of 5 to 1 (with 1 being the most dangerous).[4] Nixon also created a White House Conference on Children and Youth, which included a task force of eight students and four adults who advocated for addressing the root causes of drug abuse. And while he vastly increased law enforcement budgets and created the DEA, its stated mission was to focus on large-scale drug trafficking, not street dealing.

Perhaps the most damaging aspect of Nixon's approach was his ruthless use of selective enforcement to punish minorities and his political enemies. Twenty years after Nixon resigned, John Ehrlichman, his drug policy adviser and mastermind of the Watergate break-in, bluntly confessed to an interviewer who those enemies were and how the president pursued them: "the Nixon campaign in 1968, and the Nixon White House after that, had two enemies: black people and the anti-war left and we knew that if we could associate heroin with black people and marijuana with the hippies, we could project the police into those communities, arrest their leaders, break up their meetings and most of all, demonize them night after night

---

4    For more information on drug scheduling, see Drug Enforcement Administration, "Drugs of Abuse."

on the evening news. Did we know we were lying about the drugs? Of course, we did."[5]

All this came at a particularly bifurcated time in American public opinion. On the one hand, the increasing use of marijuana and hallucinogens had struck fear in the hearts of parents and loathing on the part of "law-abiding citizens." At the same time, there had been a significant increase in the number of Americans addicted to heroin that could, in part, be attributed to soldiers returning from Vietnam, a war that had left the country with deep mental and physical scars. Even the most law-and-order politician had a hard time sending to jail men and women whom the country had sent to fight in a losing, and perhaps totally misguided, war. In all, 20 percent of returning veterans came back addicted. Yet, remarkably, after just a year stateside, only 5 percent remained addicted,[6] a statistic that while impressive is no consolation to the many veterans who fought their own inner demons for years if not the rest of their lives—some of whom eventually ended up in jail for crimes that could legitimately be traced back to that addiction.

Gerald Ford, who became president after Nixon left the White House to avoid impeachment, also believed in tougher drug laws, including mandatory fixed sentences. But in his brief two-year term in office, Ford had trouble getting new legislation passed by a Democratic-majority Congress that still preferred prevention and treatment.

Ford's successor, Jimmy Carter, agreed with his fellow Democrats and became the first president to propose a drug budget that spent more on treatment and prevention than law enforcement.[7] Carter

---

5   Baum, "Legalize All Drugs?"
6   Clear, "How Vietnam War Veterans Broke Their Heroin Addictions."
7   Lopez, "How Obama Quietly Reshaped America's War on Drugs."

believed drug laws were too strict and even wanted to decriminalize possession of less than an ounce of marijuana, which some states had begun doing. "Penalties against possession of a drug," he said, "should not be more damaging to an individual than the use of the drug itself."[8]

At the same time, Carter was determined to craft more successful diplomatic agreements to fight cultivation and trafficking. Unfortunately, it's hard to convince nations whose gross national product depends on drugs to collaborate on effective drug-control. It's even harder to appear sincere about international drug enforcement when the CIA is making clandestine drug deals with organized crime groups in other countries.

Regardless, in 1978, Carter managed to make a deal with the Mexican government to spray their poppy fields with Agent Orange. While this successfully decreased the amount of "Mexican Mud" heroin on the market, it did not lead to the reduction in supply Carter hoped for. Instead it was a windfall for those smuggling Southeast Asian and Middle Eastern heroin. His was just one of the first of many administrations who have supported forced eradication programs despite ample proof that the supply hydra has many more heads than they can cut off and that these efforts accomplish little except to turn former friends into enemies.

President Ronald Reagan, who followed Carter, was able to put into deed things that his fellow Republicans Nixon and Ford had only been able to put into words. In terms of its severity and implicit racial prejudice, his Anti-Drug Abuse Act of 1986 could have been called the Harry Anslinger Memorial Act. Its highlights included

---

8   Carter, "Drug Abuse Message to the Congress."

mandatory minimums even for simple possession and a 100:1 disparity between the sentencing of offenses involving crack cocaine, which was considered to be favored by black users, and powder cocaine, which was favored by whites.[9] Within five years the lifetime chance an African American had of going to a state or federal prison—already significantly higher than the national average—almost doubled (from 9.3 percent to 16.5 percent), while for white Americans it rose from 2 percent to 2.5 percent.[10]

By the mid-1980s the DEA had built racial prejudice into their training regimen, encouraging law enforcement officials to use racial profiling to catch drug traffickers: "The DEA characterized certain retail and wholesale markets as controlled by racial and ethnic groups, such as Jamaicans, Haitians, Colombians, Nigerians, and Puerto Ricans. Problems soon emerged. First, the drug couriers quickly learned what the profile flags were and adjusted their methods accordingly. Second, profiling evolved without much thought given as to how to document its utility. Consequently, when questions were first raised about racial disparities in enforcement, officials had a weak empirical basis from which to defend their activities."[11]

By then, the mood in Washington had shifted so far toward fear that there seemed little room for anything else. Perhaps Reagan captured it best when he characterized the drug war in dramatic terms: "When

---

9   One of the dubious rationales for the law was the racially tinged so-called "crack baby" epidemic. South Carolina even implemented drug testing of women who were about to deliver and took their children into custody automatically if they tested positive for crack cocaine. Eventually, JAMA (*Journal of the American Medical Association*) published an article showing that there was no difference between the effect of a mother's use of crack vs. powder cocaine on her baby. See Chavkin, "Cocaine and Pregnancy."

10  Bonczan, "Prevalence of Imprisonment."

11  Riley, "Racial Profiling." Also, see Sirin, "From Nixon's War on Drugs to Obama's Drug Policies."

we say zero tolerance, we mean, simply, that we've had it. We will no longer tolerate those who sell drugs and those who buy drugs.... Those parasites who survive and even prosper by feeding off the energy and vitality and humanity of others. They must pay."[12] At the same time Reagan was making those threats, drugs had begun to play a major role in his contradictory domestic and international agendas.

Reagan was not the first president to use drugs as a weapon of foreign policy. We've already seen how the CIA used the "French Connection" to fight communists during the Cold War, even if it meant accepting the "collateral damage" of more heroin ending up in America. During the Vietnam War, the CIA doubled down on its use of drugs as a weapon of foreign policy when the charter airline company known as "Air America," which the CIA had created in the 1950s to handle covert operations, began to carry weapons to tribes in Laos in return for raw opium that was processed into heroin. Some of that heroin ended up addicting US troops; the rest was shipped to Marseilles, where the same Corsican mobs who'd played a major role in the French Connection bought it with money that the CIA could then use to buy the insurgents more weapons.[13]

During Reagan's administration the CIA tried the same strategy in Soviet-occupied Afghanistan, where it worked with one of Pakistan's intelligence services to have weapons delivered to various anti-Soviet mujahideen—in particular a group led by a man named Gulbuddin Hekmatyar, who was willing to supply truckloads of opium and heroin in exchange for weapons. The administration tried a similar policy

12  Reagan, "Reagan Radio Address to the Nation."
13  See Chouvy, *Opium*, for a comprehensive look at the geopolitical impact of the opium poppy, particularly in modern times, as well as the many roles the CIA's Air America played in transporting everything from spies and drug enforcement officers to illicit cargo throughout Southeast Asia.

in Nicaragua, when it decided to help a group in Nicaragua known as the Contras (whom Reagan referred to as "freedom fighters") that was trying to overthrow the Marxist-leaning Sandinista regime. In what became a notorious scandal, the administration sanctioned a secret deal in which the United States traded missiles and other arms to its enemy Iran in exchange for Iran persuading Lebanon to free some Americans held hostage there. The money left over from the deal was used to support the Contras, who, it was discovered, were well connected with the smuggling rings that brought drugs into America—primarily cocaine, which was America's drug of choice at the time.[14]

The CIA's justification was that, while the country considered drugs to be bad, communism was way worse and, if the lives of "a few" citizens and soldiers had to be sacrificed, along with any semblance of geopolitical ethics, one simply had to look the other way—even if "one" was the president or vice president of the United States.

Indeed, when Vice President George H. W. Bush succeeded Reagan in office, he famously claimed that he was "out of the loop" on the Iran-Contra deal even though, as a former director of the CIA, he undoubtedly knew where the bodies were buried and who had put them there.[15] Bush ramped up Reagan's rhetoric, saying, "If you sell drugs, you will be caught," although he clearly wasn't referring to his own drug-dealing CIA agents. "And once you're convicted, you will do time," he added, although he clearly didn't mean Oliver North, who orchestrated the whole scheme and didn't serve a day in

---

14   The administration had to keep the deal secret because Congress had passed a law that prevented the Reagan administration from directly funding the Contras.
15   "Bush 'Out of the Loop' on Iran-Contra?" Another important player in the Iran-Contra affair was Panama's dictator Manuel Noriega, whom then Vice President Bush met with as late as 1983. Gerth and Times, "Bush and Noriega."

prison. Bush insisted that "we need more prisons, more jails, more courts, more prosecutors," and threatened to deny federal funds to any school, college, or workplace that didn't have tough drug policies in place. "Period!" Bush concluded.[16] This travesty of injustice was taking place at the same time ordinary citizens were being caught with small quantities of drugs for personal use and sentenced to lengthy prison terms.

Bush backed up his rhetoric with dollars, proposing a $2.2 billion increase in drug spending—only 4 percent of which was for treatment, a percentage so outrageously low that Congress insisted he add another $1.1 billion for treatment, education, and prevention. Even with that, the *New York Times* called his efforts to reduce drug trafficking and sales a dismal failure that represented "new funds but an old strategy."[17]

President Bush's successor, Bill Clinton, quickly proved he could be as tough on crime as any Republican. He not only continued Bush's policies but initiated ones that proved even more damaging in terms of sentencing guidelines. On Sep 13, 1994, he signed the largest crime bill in US history: the Violent Crime Control and Law Enforcement Act, which expanded federal drug sentences and reintroduced the possibility of the death penalty.

During his first four years, Clinton managed to spend more on anti-drug efforts than Reagan and Bush spent in their twelve years in office combined, including a particularly duplicitous billion dollars to buy "advertising" time. Instead of buying ads per se, the Office of National Drug Control Policy (ONDCP) asked the networks to "embed" anti-drug messages in their shows, in the same way that companies pay for

16    "Presidential Address on National Drug Policy."
17    Treaster, "Four Years of Bush's Drug War."

product placement, a tactic troublingly reminiscent of the government brainwashing techniques George Orwell predicted in *1984*.

The ONDCP actually used sophisticated analysis to determine how much a subliminal message was worth in a particular show. Even more controversially, it reviewed scripts and "negotiated" changes, but as Clinton and the presidents who followed him would discover, no analysis could prepare America for the drug crisis to come, one that began not in China or Turkey or Afghanistan or Mexico—but at a small pharmaceutical company in Stamford, Connecticut.[18]

\* \* \*

In 1995, two brothers, Raymond and Mortimer Sackler, were made honorary Knight Commanders of the Order of the British Empire by Queen Elizabeth II in honor of their "professional, humanitarian and exploration" achievements—an award that has been bestowed on an eclectic assortment of luminaries including Mother Theresa, Bono, and J. Edgar Hoover.

The same year, Purdue Pharma, a company owned by the Sacklers, received approval from the FDA for OxyContin, a new extended-release painkiller that would later be blamed for starting an epidemic of opioid addiction and overdoses that America and many other countries continue to spend vast resources trying to contain today.

Since then, Purdue has been sued, vilified, and held up as a poster child for everything wrong with the healthcare system in general and pharmaceutical companies in particular.

The research scientists at Purdue who developed OxyContin made

---

18   Forbes, "Prime-Time Propaganda."

the same mistake Frederick Sertürner had made when he developed morphine in 1803; the same mistake Heinrich Dreser at the Bayer Company had made when he synthesized heroin in 1895; the same mistake Martin Freund and Edmund Speyer had made in 1916 when they synthesized oxycodone; the same mistake Otto Eisleb and Otto Schaumann had made in 1939 when they synthesized meperidine (Demerol); and the same mistake Carl Mannich and Helene Lowenheim had made when they synthesized hydrocodone in 1943.[19]

All of them believed they were developing less addictive and more effective alternatives to existing opioids. All of them were wrong.

In terms of marketing and sales, the timing of OxyContin's release could not have been better. In the late 1990s, the VA hospital system began to consider pain the fifth vital sign (along with pulse, temperature, respiration, and blood pressure). Doctors were encouraged to deal with pain more aggressively, which meant walking an even finer line between providing relief and risking addiction.[20] Purdue argued that, as an extended release version of oxycodone (which had been a standard opioid medication in America since 1939), OxyContin would be less addictive because it delivered lower doses of the drug over longer time periods. They, too, were wrong.

Patients had become addicted to prescription opioids such as oxycodone and hydrocodone throughout the twentieth century as well as barbiturates (e.g., Seconal and Nembutal) and benzodiazepines

---

19  Morphine is derived directly from the opium sap; hydrocodone is a semi-synthetic derived from codeine; and oxycodone is derived from thebaine (the third opium alkaloid). While all are used for severe pain, they differ in terms of side effects and mode of action. Vicodin is the best-known brand name of hydrocodone. Percocet is oxycodone with acetaminophen. Percodan is oxycodone with aspirin. Hydrocodone tends to be used more often for severe coughs. See Wasacz, "Natural and Synthetic Narcotic Drugs."

20  Morone and Weiner, "Pain as the 5th Vital Sign."

(e.g., Valium). But the numbers were minimal compared to the epidemic of opioid addiction that began when doctors began over-prescribing OxyContin and Purdue began concealing its risk.[21]

That risk was even greater when it was tampered with—e.g., by crushing and snorting or dissolving and injecting. Both techniques brought a sense of deep relaxation and euphoria similar to heroin—at least the first time. Increasingly, patients who had begun taking Oxy-Contin legitimately for pain relief found themselves chasing this high.

Regardless of their path to addiction, when a patient could no longer get any by prescription—or simply wanted more than they were prescribed—they would buy some from friends and then strangers on the street. And when that supply dried up, many turned to more readily available and less expensive heroin.

Doctors and patients wanted OxyContin. The FDA approved it. The government's efforts to mandate data monitoring and sharing technologies that would expose prescription drug fraud were woefully inadequate. Was it really Purdue Pharma's fault it was so widely used and abused? After all, what they did was simply the continuation of a long tradition of poorly tested medical miracles and shameless overmarketing. Were they really so much worse than "Mrs. Winslow," who in the 1800s claimed right on the label of Mrs. Winslow's Soothing Syrup that it was *The Mother's Friend for Children Teething*? (Unbeknownst to many mothers—and the person on the label who looks suspiciously like a nanny—the bottle had 65 mg of morphine in it.)[22] Similarly, when Bayer introduced heroin, it claimed the new

---

21   See Armstrong, "Sackler Embraced Plan to Conceal OxyContin's Strength From Doctors, Sealed Testimony Shows."

22   There are many variables in recommended morphine dosage, but it can be as low as under 1 mg for a child. Pain Assessment and Management Initiative, "Pain Management and Dosing Guide."

drug was the "Cheapest Specific for the Relief of Coughs (bronchitis, phthisis whooping cough etc.)."[23]

Actually, it *was* Purdue's fault. If they had just been self-deceiving hucksters like those patent remedy marketers of the 1800s, they—as well as the FDA and the doctors who trusted them—would "simply" be guilty of sloppy science and wishful thinking. Their crime was ignoring and concealing indisputable reports from the field that the pills were being overprescribed, sold on the street, and adulterated.[24] By the time the company came out with an "abuse-deterrent formulation" (which, ultimately, didn't completely deter abuse), and states slowly began to pressure doctors (and pharmacies) to decrease prescriptions, many users had already turned to heroin.[25]

In 2007, the president, top lawyer, and former chief medical officer of Purdue Pharma were ordered to pay $634.5 million in fines. The company itself agreed to pay $19.5 million to twenty-six states and the District of Columbia (Purdue earned over $1 billion in OxyContin sales that year). Ten years later, in September 2018, as part of negotiations to settle 1,000 lawsuits, Purdue reportedly offered to give away buprenorphine—one of the main drugs used in medication-assisted treatment for addiction. The patent for buprenorphine is owned by a company known as Rhodes Pharmaceuticals, which is owned by the Sackler family.[26]

Estimates are that 200,000 Americans have died directly from OxyContin-related overdoses since 1999. Countless other addicts have been given long, if not lifetime, sentences for opioid sales and use.

---

23  "Drug Ads Gallery, 1900–1909."
24  Meier, "In Guilty Plea, OxyContin Maker to Pay $600 Million."
25  Meier, "Origins of an Epidemic."
26  Terry, "In Negotiations of 1,000 Opioid Lawsuits, Purdue Pharma Offers Free Opioid Therapy."

By the end of 2017, Purdue had sold approximately $35 *billion* worth of OxyContin.[27]

No one from Purdue has ever served time in prison.

\* \* \*

While people were just becoming aware of how dangerous OxyContin was, George W. Bush became president. At first, it appeared his drug policy would be more compassionate than Clinton's. While there was $2.3 billion for border control in his first budget, it also included $3.8 billion for treatment and research as well as $650 million for youth education programs at schools and in the community.

Yet, however sincere his attempt to deal with the drug problem in a more enlightened way, his response to 9/11 would end up exacerbating it significantly, as once again, America made opium a foreign policy tool.

Afghanistan's opium production had increased by twenty times between the 1970s and the early 1990s, thanks in part to the CIA using the drug to fund Afghanistan's resistance to Soviet occupation. Once the Soviets left Afghanistan, the country plunged into civil war until the Taliban, backed by al-Qaeda, formed a tenuous central government—one whose gross national product depended heavily on raw opium and heroin. While the West officially frowned on this, it was too busy in the Middle East by then to address it.

By 2000, the Taliban controlled about 75 percent of the country, with the rest controlled by the Northern Alliance, which was supported by Russia and other neighboring countries. When a severe

---

27   Morrell, "The OxyContin Clan."

drought destroyed the poppy crop, the Taliban government needed money badly and looked to the United Nations for foreign aid, offering in exchange to crack down on future opium production. As far as the international community was concerned, Afghanistan had to stop supporting terrorism and violating human rights before they would be given any assistance. Bush, seeing the opportunity to expand American influence in the region, agreed to give them $43 million in foreign aid—a decision he soon regretted when Osama bin Laden (whom the Taliban had been protecting on behalf of his Islamic terrorist group al-Qaeda) choreographed the most successful terrorist attack on US soil in history.

After 9/11, when the Taliban refused to turn over bin Laden, the United States and Great Britain attacked alongside their former enemies, the Northern Alliance. The Taliban government crumbled and a new government was formally elected. The Taliban started a counterinsurgency movement, which they financed by getting back in the opium business. So, in turn, President George Bush decided to get back into the opium-*fighting* business and authorized major crop-eradication efforts. He also turned a blind eye to certain factional warlords of the Northern Alliance who were allied with the United States versus the Taliban, despite abundant evidence that they, too, were involved in heroin trafficking.[28]

Crop eradication is a zero-sum game. There are 10,000 seeds in a single opium poppy pod. A farmer whose crop has been destroyed (or who has been driven off his or her land) can easily walk away with

28 In 2005, the *New Yorker* reported: "General Abdul Rashid Dostum...factional warlord of Mazzar I Sharriff—was the biggest heroin trafficker out of Afghanistan and became our ally in the Iraq and Afghani Wars....Meanwhile, General Abdul Rashid Dostum...an Uzbek who dominates a large swath of northern Afghanistan, is viewed by most human-rights organizations as among the worst war criminals in the country." Anderson, "The Man in the Palace."

enough seeds to start over. Not only do new crops and other sources of supply seem to pop up overnight, eradication turns farmers who are just trying to feed and shelter their families into willing recruits for governments, terrorist organizations, and revolutionary groups. As one Afghan woman shouted, "They will have to roll over me and kill me before they can kill my poppy."[29]

If farmers could make as much money (or at least a reasonable living) growing legal crops and raising flocks of sheep, goats, and cows, most would be happy to do so. It would certainly be far less dangerous than being in the middle of warring factions in the drug trade. Instead, after years of warfare, Afghanistan's irrigation systems have been ruined, orchards devastated, animal flocks decimated, and seed supplies destroyed. The United States has spent billions of dollars on military solutions, money that would be far better spent "winning the hearts and minds of the people" by rebuilding the country's economic infrastructure.[30]

This is obvious to some of those who have witnessed the problem firsthand. It is the reason that a group of US military veterans started a nonprofit called Rumi Spice to help farmers in Wardak Province grow and market premium saffron, one of the world's most expensive spices.[31] At approximately $200 per kilogram, it doesn't quite match the $300 per kilogram they could make from poppies but it's much more reliable, sustainable, and safe to grow; and since, like poppy cultivation, it's labor intensive, it employs a similar number of people. The country's agricultural ministry has embarked on a similar initiative in the province of Herat, where there are now 400 farmers in the saffron business.[32]

In the long run, crop replacement might do no more than crop

29   McCoy, "How the Heroin Trade Explains the US-UK Failure in Afghanistan."
30   Ibid.
31   "Rumi Spice: About Us."
32   Ghafour, "Afghan Farmers Turn to Saffron."

eradication to reduce overall worldwide supply, since other countries can increase their crops within a season to compensate. However, it is certainly a more effective and honorable—and potentially even more strategically successful—way to work with countries that have been ravaged by drugs, in part due to America's nation-building efforts.

When Barack Obama came into office, he renewed America's commitment to defeat the Taliban and, by the end of 2009, he had raised the number of US forces to 100,000 as part of what he called a "surge" to make it possible to begin *withdrawing* troops. After the assassination of Osama bin Laden, Congress began to push for more rapid troop withdrawal. By the end of 2015, there were only about 10,000 American troops still in the country.[33]

Some foreign policy experts continue to insist that destroying poppy crops is essential to accomplish our goals in Afghanistan. They also say troop levels will have to go up again to 100,000 or more to do that job. They point out that, when the number of soldiers tripled (from 30,000 to more than 90,000) between 2007 and 2012, the area under cultivation went down 25 percent (200,000 hectares to 150,000 hectares). Whereas after Obama's troop reductions the hectares under production more than doubled.[34]

If only the math were so simple. If only more troops = fewer poppies = 0 Taliban = America wins. However, other experts argue that, no matter how much nation-building and investing the West does to support Afghanistan's economy, it won't have a major impact

---

33    See Council on Foreign Relations, "A Timeline of the U.S. War in Afghanistan"; and Associated Press, "A Timeline of U.S. Troop Levels in Afghanistan since 2001."

34    Islamic Republic of Afghanistan Ministry of Counter Narcotics, "Afghanistan Opium Survey 2017."

on production because, for most of its people, drugs remain the most reliable source of income.[35]

One writer, after enumerating the total number of troops deployed, soldiers killed, and stunning amount of money spent in Afghanistan from 2001 to 2017, concluded, "In the American failure lies a paradox: Washington's massive military juggernaut has been stopped in its steel tracks by a small pink flower—the opium poppy."[36]

Even if America "succeeded" in Afghanistan there's one very inconvenient truth that changes all these equations. Less than 10 percent of the heroin smuggled into America originates there. Most of the heroin entering the country is instead processed from poppies grown in Southeast Asia and increasingly Mexico and other Latin American countries.[37] In other words, even if there weren't a single poppy growing in Afghanistan, America would still be awash in opiates.

The even more inconvenient truth is that heroin's long reign as the primary addictive "hard" street opiate is coming to an end, *regardless* where it's being grown. Free from the ingredients within the opium flower, opiate analogs are synthesized rather than grown and are designed to be hundreds of times stronger than heroin.

\* \* \*

Heroin isn't the only drug coming from Mexico—their cartels are also industry leaders in the most powerful factor in America's opium crisis today: lethal synthetic opioids that are often used to cut heroin, making it multiple-times deadlier.

---

35  Felbab-Brown, "Afghanistan's Opium Production Is through the Roof."
36  McCoy, "How the Heroin Trade Explains the US-UK Failure in Afghanistan."
37  Isacson, "Four Common Misconceptions about U.S.-Bound Drug Flows."

While headlines give the impression that fentanyl, the most famous synthetic opioid, appeared out of nowhere in the last decade or so, it was actually first synthesized back in 1959. When used in a patch form, it releases a slow, remarkably effective dose of the drug. As prescribed, it is frequently and safely used for pre-operative anesthesia, and has spared thousands of cancer and other terminal patients from suffering end-of-life pain. As contraband, however, it has killed thousands.

In other words, the impression we have of Big Pharma scientists shamelessly coming up with increasingly addictive potentially deadly drugs without any legitimate medical use simply doesn't tell the whole story. Rather, it has taken a perfect storm of overmarketing and overprescribing combined with the ease of adulteration and the ingenuity of drug cartels to create today's crisis.

In addition, there's another reason the opioid crisis keeps growing: with a certain amount of chemistry expertise, it's possible for anyone to take readily available chemicals, synthesize them into fentanyl, and ship the resulting drug directly from a lab to the street with only a middleman or two in between.

Finding the recipe would be easy. Its developer, Dr. Paul Janssen, patented it decades ago (United States of America patent 3164600). So it's in the public domain. The process simply involves: "condensing propionyl chloride with N-(4-piperidyl) aniline, then treating the resulting N-(4-piperidyl) propionanilide with phenethyl chloride, aiding each condensation by the presence of a suitable dehydrochlorinating agent. Reaction of the base with an equimolar portion of citric acid yields the (1:1) citrate. Fentanyl citrate."[38] The process takes about a week. And the result looks and cooks like heroin.

---

38  "TOXNET."

The most famous rogue producer of opioids was George Marquardt, a brilliant, self-taught "mad scientist" who proved during the 1980s that drugs can be manufactured virtually anywhere from readily available ingredients. (If his strange brews hadn't been so lethal, he would have been the Paracelsus of our time.) Marquardt's main "achievement" was figuring out how to manufacture fentanyl (along with hallucinogens and methamphetamines) in the privacy of his own lab. He even made the precursor chemicals himself and measured their purity with a home-built mass spectrometer. After making millions of dollars, he was caught in the early 1990s. When asked his profession, he told the judge, "Drug manufacturer." When asked what kind, he answered, "Clandestine." (Which does sound exactly like something Paracelsus would say.) The judge sentenced him to twenty-five years. He served twenty-two and died in 2017, a year after being released.

Reporters described him as a "mythical figure," an "evil genius," and a "serial killer." He certainly didn't have much of a conscience. When the police finally found him and raided his Oklahoma lab, he "walked them through his process for cooking meth and quickly pled guilty to his charges."[39]

Illegal drug manufacturers have also been able to take advantage of another loophole that the governments of the United States and China, in particular, have been trying to close. Just as drugs aren't dangerous until people use them dangerously, it's equally true—and often equally overlooked—that they aren't illegal until a country criminalizes them. In America, it's up to the DEA (in consultation with the FDA) to decide which drugs to put on which of the schedules that it started using during Nixon's administration; state and federal

---

39   Moskowitze and the Fusion Investigative Team, "Last Chances in the Second City."

regulations rely on this for minimum sentences. This means there's a time lag between the time a drug shows up on the street and when the DEA schedules it.[40]

To further complicate things, since it's difficult to outlaw a drug that hasn't been invented yet, all it takes is minor tinkering with an existing formula to develop an opioid-like drug (i.e., any drug that triggers opioid receptors) that's equally "effective." These new drugs are called "analogues." It's as if heroin were a shape-shifting criminal. In parts of the United States today, street "heroin" should be presumed to include fentanyl.

After spending a few years scrambling to catch up, the government acknowledged the challenge of keeping abreast of drug innovations and, claiming an emergency measure was "necessary to avoid an imminent hazard to the public safety," declared all "Fentanyl Related Substances" Schedule 1 drugs for two years. By "fentanyl-related," the DEA included its "isomers, esters, ethers, salts and salts of isomers, esters, and ethers." The ruling criminalizes people "who handle (manufacture, distribute, reverse distribute, import, export, engage in research, conduct instructional activities or chemical analysis, or possess), or propose to handle fentanyl-related substances."[41]

It may seem counterintuitive that it would be profitable for heroin manufacturers to go to the extra step (and risk) of cutting their heroin with a mere "dusting" of deadly fentanyl. The reason lies in the economics of its production. Making heroin involves hiring or outsourcing armies of workers to plant acres of poppies, carefully collect

---

40   Such lag times might be minimal in the face of an acute crisis or quite extensive—but, for example, academics and the CIA were experimenting with LSD in the 1950s, and it wasn't illegal until 1968.
41   "2018—Temporary Scheduling Order: Temporary Placement of Fentanyl-Related Substances in Schedule I."

bucket loads of sap, and refine it into heroin—and then establishing a sophisticated smuggling operation that involves multiple hand-offs and markups—before it reaches the user. Random events such as wars and droughts can devastate an entire season's crop. Drug cartels have avoided this entire process by sending chemists to America along with the legal precursor chemicals he or she needs to turn them into illegal drugs that hit the street almost immediately, giving authorities minimal opportunity to intercept the supply. In the end, $1,000 worth of wholesale heroin can generate a few thousand dollars profit. The profit on the same number of doses of fentanyl could be close to $8 million.[42]

The person often credited with first making a large-scale killing in the heroin-fentanyl business is Mexico's most famous drug lord, a second-grade dropout named Joaquín "El Chapo" Guzmán, who, having escaped twice from maximum security prisons in Mexico, was held in virtual seclusion at the Manhattan Correctional Center, a few blocks from City Hall, during his trial in Brooklyn for an ongoing criminal enterprise, including murder, money-laundering, and use of firearms. Security was so tight that every time he was brought to a hearing, the Brooklyn Bridge had to be shut down to make way for a police motorcade that included an ambulance and SWAT team.

On February 12, 2019, a jury found him guilty on all ten counts of the indictment. If the verdict is upheld on appeals, he will be sentenced to life in prison.[43]

A cross between Al Capone and Caligula, El Chapo's life story is a remarkable tale of the rise (and apparent fall) of a major drug lord.

---

42   Deprez, Hui, and Wills, "Deadly Chinese Fentanyl." Also see Quinones, *Dreamland*, which describes in depth how opioids spread through suburban and small-town America.
43   Feuer, "El Chapo Found Guilty."

After spending his errant youth in a town with no electricity, El Chapo built an empire that, for several decades, left a trail of indiscriminately slaughtered allies and enemies as well as frustrated drug agents—all while he moved around Mexico in bulletproof cars, indulged his tastes in fine dining and beautiful women, and allegedly bribed people at "nearly every level of the Mexican police, military and political establishments."[44] The two times he was arrested in Mexico, he allegedly bribed his jailers (as well as the generals and governors he'd been bribing all along) and continued running his empire without missing a beat until his confederates finished digging the tunnels through which he could make his escape.[45]

These escapades have made him a cultural icon in the classic "outlaw" mold. (Indeed, his last interview before being caught was with Sean Penn for *Rolling Stone;* Penn, while clearly depicting him as a debauched murderous thug, had to acknowledge the romantic appeal of this "rags-to-riches" master criminal.) DEA Special Agent Jack Riley, who played a major role in El Chapo's capture, put a far more sobering spin on the story: "All those routes he opened, all that fentanyl he shipped—he's gonna kill our kids for years to come. This monster he built....It's too big to fail now, thanks to him."[46] As he and his lawyers fought the litany of charges that had been leveled against him, El Chapo's family, former business partners, and other cartel leaders weren't missing a beat.[47] Indeed, in El Chapo's absence, the Mexican drug trade has splintered into dozens of viciously competitive crime groups, vying for the kind of leadership he held for years, while heroin

---

44    Ibid.
45    Penn, "El Chapo Speaks."
46    Quoted in Solotaroff, "El Chapo."
47    See, for example, Holman, "El Chapo Trial"; and Eustachewich, "El Chapo's Mexico Drug Cartel Is Doing Just Fine without Him."

and fentanyl continue to flow into America from Mexico via tunnels, boats, trains, planes, donkeys, and couriers.[48] In fact, according to the *New York Times*, Mexican heroin production increased by 37 percent and fentanyl seizures at the border more than doubled *after* El Chapo was arrested.[49]

Another significant loophole that illicit opioid manufacturers and dealers have been using involves producing drugs in other countries and using the Internet to market them. For several years, Chinese chemists were formulating fentanyl and two dozen other drugs that were then perfectly legal there. While they tried to be careful to make sure they stayed a step ahead of their own country's drug regulations— and would profess ignorance of the ultimate use of their concoctions— their shipping practices, which involved stuffing the drugs in packages labeled as clothing, buttons, and radios, imply that they knew what they were sending would not be welcomed by US Customs.[50]

In response, the Chinese government began cooperating with US authorities to stop the manufacture of opioids in their country. In the last few years, they have regulated 130 synthetics including ten fentanyl analogs, and shut down companies manufacturing them. Still, as the Chinese are perhaps all too aware, smuggling can never be extinguished as long as there is a market for the drug. As Yu Haibin of China's National Narcotics Control Commission put it, "The biggest difficulty China faces in opioid control is that such drugs

---

48   Smuggling also ensnares apparently innocent civilians like the eighty-one-year-old woman who was arrested for trying to smuggle almost a million dollars of heroin into San Diego. Riggins, "2 Women Accused of Smuggling Arrested at Border."

49   Feuer, "El Chapo Found Guilty."

50   At the same time, they accepted payments by credit card and PayPal, and provided next-day delivery and 100 percent money-back guarantees, which implies that they weren't afraid of getting caught for dealing, just of having their shipments confiscated.

are in enormous demand in the U.S."[51] Indeed, despite the bellicose claims of many American officials, William Brownfield, former assistant secretary of state for the Bureau of International Narcotics and Law Enforcement Affairs, acknowledged in 2018 that relations with China had improved "astronomically" and that the country had done "gargantuan work" controlling products and, in the process, saved many American lives. In spite of this, and his insistence that relations with Mexico when it comes to combating drug production and trafficking were at historically high levels, opioid consumption in America is higher than ever.[52]

There's yet one other supply line that authorities have had trouble shutting down: the "Dark Web." Originally developed by US Navy mathematicians to create a secure area on the Internet, it now refers to anything stored online that won't appear in typical searches. Accessing the Dark Web requires using a specialized browser, the most famous of which is called TOR, and while using it is perfectly legal, visiting sites that are doing illegal business on it is not. Since the nature of the Dark Web also makes it easy to hide one's identity from all but the most determined authorities when using it, it has, with the help of untraceable crypto-currencies like Bitcoin, become an active marketplace for passports, poisons, pornography, and just about anything else that can get past postal service screeners or be delivered digitally.[53]

The first drug "kingpin" on the Dark Web was a now-thirty-

---

51  Deprez, Hui, and Wills, "Deadly Chinese Fentanyl."
52  Brownfield, "Briefing on the International Narcotics Control Strategy Report."
53  There are many descriptions of how the Dark Web works that can be accessed using good old-fashioned browsers. For a good basic description see Jacobson, "Silk Road 101." There are also many different descriptions of bitcoins and the associated blockchain. But it takes a lot more than regular linear thinking to understand them.

something libertarian computer programmer named Ross Ulbricht, who started a site called the Silk Road where more than a billion dollars of goods changed hands. While as of this writing Ulbricht is spending life in prison without parole, he is yet another example of how, in the race between drug dealing and law enforcement, the latter is usually one step behind the curve until the dealer makes a false move.[54] Ulbricht himself made several such false moves, which is the reason he's in jail while the creators of similar Dark Web marketplaces are not, but before he was caught, he demonstrated the never-ending ingenuity of dealers to thwart government assaults on opioids.[55]

While Ulbricht serves his life sentence in Colorado, drug sales on the Dark Web continue unabated. In fact, they doubled after his trial, as all the news coverage raised awareness of the marketplace.[56]

54   There's a move to have him pardoned. See Whigham, "Silk Road Founder Ross Ulbricht Sends Messages from Prison."
55   Hern, "Five Stupid Things Dread Pirate Roberts Did to Get Arrested."
56   Greenberg, "Silk Road Creator Ross Ulbricht Loses His Life Sentence Appeal."

# Chapter 28

# The $1 Trillion Question: What Do We Do Now?[1]

We cannot talk in absolutes, that drug abuse will cease, that no more illegal drugs will cross our borders—because if we are honest with ourselves we know that's beyond our power.

—President Jimmy Carter

We can be the generation that ends the opioid epidemic. We can do it.

—President Donald Trump

A world where all people live free of the burden of drug abuse.

—Vision of Drug-Free America Foundation

Unfortunately, if history is any guide, Carter is right, Trump is wrong, and the Foundation's vision, while admirable, is not realistic. As the

---

1   Comprehensive reports on dealing with the opioid crisis include Bollinger and Burns, "Ending the Opioid Crisis"; The President's Commission, "On Combating Drug Addiction and the Opioid Crisis"; and Kennedy, "Recommendations of Congressman Patrick J. Kennedy." The figure of $1 trillion for the war on drugs is cited in several sources including Coyne and Hall, "Four Decades and Counting."

long history of opium illustrates, most approaches to controlling the drug trade have proven ineffective. Fortunately, there *are* an increasing number of effective prevention strategies, as well as treatments that can reduce harm and help those who are addicted.

But first, let's be clear: "drug wars" don't work. That's been demonstrated by everyone who has declared one, from the seventeenth-century Chinese emperor Yongzheng to Richard Nixon and every American president since then.[2, 3] Whether a country fights the drug war on the growing fields and in the labs of foreign countries or takes a criminal-justice approach that involves waging war on its own citizens, the history of drug wars is one of failed promises and devastating human casualty.

The United States is currently monitoring drug activities in almost seventy countries—and coordinating interdiction efforts with dozens of them. Still, processors, smugglers, and dealers—like squirrels laughing at squirrel repellent—always find a way. To make things worse, America has not only failed to achieve lasting positive results but, at times, has acted contrary to its own professed drug policies and formed alliances with alleged enemies because of its convoluted international political priorities.

Dealing effectively with the current crisis requires the same FEMA-like urgency and massive funding that we'd use in response to any national disaster.

In October 2018, it appeared that Congress and President Donald Trump were taking the first step in that direction when they enacted the Substance Use-Disorder Prevention that Promotes

---

2   Even though, in the 1950s, China waged the drug war to end all drug wars—and *still* has severe penalties for drug possession and use—there are an estimated two million addicts and another 10 million users in the country today. See Griner, "China's New Opium Wars."

3   Except Barack Obama who dropped the word "war" from his own administration's battles to fight drugs in America and abroad.

Opioid Recovery and Treatment for Patients and Communities Act (SUPPORT). It was an uncommon display of bipartisanship, reflecting how the opioid epidemic has spread into every voting district in the country.

The law is hundreds of pages long, with nine chapters and 115 sections that cover an extensive array of initiatives to stop opioid abuse: it authorizes Medicaid and Medicare to cover more abuse-related screening and treatment programs; it orders the FDA to clarify its regulations regarding pain products; it gives customs officials more power to identify and confiscate illegal imports; and it gives healthcare professionals more flexibility to provide medication-assisted treatment (MAT). There are also regulations intended to reduce the spread of prescription opioids, by mandating better drug labeling and authorizing pharmacists to deny suspicious prescriptions. Yet many observers consider this law to be worth little more than the paper of the *Congressional Record* it's printed on, because it authorizes *only* $8.5 billion for *only* one year (an amount that includes the costs already budgeted for Medicare and Medicaid drug treatment). That may sound like a significant investment in the war on drugs but it is a far cry (and many unnecessary deaths) from a FEMA-like response, since, in 2017, the government spent fifteen times as much—a total of more than $130 billion—on disaster relief.[4]

Under the circumstances, it's no surprise that experts and activists recommend a far more comprehensive and ambitious response—in particular a major expansion in addiction treatment.[5] Arguing that all the bill does is double down on strategies that have, so far, only scratched the surface of the problem, Dr. Jeffrey Singer, a senior fellow

---

4  Lingle, Kousky, and Shabman, "Federal Disaster Rebuilding Spending."
5  Lopez, "Trump Just Signed a Bipartisan Bill to Confront the Opioid Epidemic."

at the Cato Institute, wrote, "Sadly, all that Congress and the White House have to brag about is a policy that is driving non-medical users to more dangerous drugs and causing desperate pain patients to turn to the black market or to suicide for relief."[6] Similarly, former congressman Patrick Kennedy wrote: "I hope Congress doesn't think they can put this behind them because they passed these bills. It takes an urgency like we had during HIV-AIDS.... It takes political will."[7]

\* \* \*

### CRIMINAL JUSTICE REFORM

In spite of ample evidence to the contrary, politicians continue to believe that doing away with drugs is as simple as outlawing them.

Every effective strategy designed to reduce the number of people selling, using, and dying from opioids involves accepting that addiction needs to be addressed as a disease, *not* a crime. The way America deals with tobacco and alcohol (which are equally lethal over the long term) is a reasonable starting point for doing that.

Once researchers began making the connection between smoking and lung disease in the 1950s, the government began a process of regulation, education, and taxation, including laws against sales to minors, laws forbidding cigarette advertising and smoking in many public places, and anti-smoking education programs funded by "sin" taxes on cigarettes. These strategies have significantly reduced the number of people who smoke, and each could be adapted to deal with the opium crisis by raising funds for education and prevention while

---

6   Singer, "The Administration's Fundamental Flaw on Opioid Addiction."
7   Quoted in Itkowitz, "Senate Easily Passes Sweeping Opioids Legislation."

reducing violent crime (just as the end of Prohibition reduced the violent mob-controlled alcohol trade of the 1920s).

Mandatory sentences just make the problem worse. They use valuable government resources to incarcerate people who don't pose a physical threat to anyone but themselves and whose addiction is likely to become more serious in the process because, while the government would never consider withholding medical treatment from a prisoner suffering from heart disease or cancer, many prisoners do not have access to MAT in prisons.[8] So, addicts are forced to go cold turkey (which has a low rate of success) or resort to an illicit prison drug trade that can lead to even longer periods of confinement. Also, like all prisoners, after release they often find it extremely difficult to find housing and employment—stresses that increase the risk of relapse.[9] In the same regard, expunging the criminal records of those convicted of nonviolent, no-distribution crimes would improve their chances of success when they reenter society.

The most innovative and comprehensive criminal justice reform program is the "Portuguese Model," which, while radical and far from perfect, has led to significant gains in that country's fight against addiction. The experiment began in 2001, when the nation decriminalized the possession and consumption of all illegal substances. While dealers are still prosecuted, users caught with up to a ten-day supply are given a warning, a fine, or directed toward treatment, harm reduction, and support services. Today, the number of drug-related deaths in

---

8   In early May 2019, the First Circuit Court in Boston became the first federal appeals court to rule that addicts had the right to receive Medication Assisted Treatment in jail. The ruling, which was based on the American Disabilities Act, is a major step toward protecting the rights of inmates to be treated for mental as well as physical illnesses. See Arnold "Setting Precedent, a Federal Court Rules Jail Must give Inmate Addiction Treatment."

9   "Why Imprisonment Is More Harm Than Help to Addicted Offenders."

Portugal is just 10 percent the rate in Great Britain and a remarkable 2 percent of that in the United States.[10]

The effectiveness of the program may be partly due to decriminalization itself, but as important and innovative is the unconditional acceptance of addiction as a disorder rather than a moral failing, which reduces the stigma involved in seeking treatment. Plus, instead of trying to treat addiction in isolation, healthcare workers have the resources to address the wide range of physical and emotional problems that usually accompany the disease.

Furthermore, the program implements mechanisms for spreading the word about its services: instead of expecting addicts to come to them. For example, public health officials in some Portuguese cities cruise the streets providing free methadone to users willing to work toward rehabilitation, and safe needles to those who are not.

While the Portuguese government accepts that completely ending drug use is impossible and advocates believe it could still be doing more, millions of dollars that would have gone to interdiction and enforcement are invested in public health services, with impressive results.

* * *

Though less ambitious than the Portuguese legalization model or full decriminalization, drug courts, which provide nonviolent offenders the option of going into treatment instead of prison, have shown some success. Since the creation of the first drug court in Miami-Dade County in 1989, they have been established or are being planned

---

10   See Ferreira, "Portugal's Radical Drugs Policy Is Working" and Kristof, "How to Win a War on Drugs."

in all fifty states and there are a host of studies demonstrating their effectiveness at reducing recidivism.[11]

Still a wide range of factors affect their success, including the skills of the judge, the level of assessment and treatment provided, court staff turnover, and resources provided to the program. There are also professional, medical, and ethical issues involved. While a judge's ability to make a meaningful connection with an offender can have a major effect on the person's willingness to participate in the program, they usually do not have the medical expertise required to evaluate what type of treatment(s) might prove most effective.[12]

There is also always the danger of programs that bring together users serious about recovery with those who are *compelled* to undergo it, as well as questions regarding whether success rates are due, in part, to "cherry-picking" those who have been charged with lesser crimes. Finally, there are subtle human rights issues in terms of the right to privacy and autonomy that the accused has to relinquish to participate. All of these issues, along with government analyses, indicate that drug courts may not be as effective as they seem. As one study concluded: "Evidence does not, on the whole, suggest improved outcomes related to compulsory treatment approaches, with some studies suggesting potential harms. Given the potential for human rights abuses within compulsory treatment settings, non-compulsory treatment modalities should be prioritized by policymakers seeking to reduce drug-related harms."[13]

---

11   Office of Justice Programs, "Do Drug Courts Work?" Also see "Drug Treatment Court Database."
12   Charles, "Criminal Justice Reformers Are Hooked on Drug Courts."
13   Werb et al., "The Effectiveness of Compulsory Drug Treatment."

## SUPPLY REDUCTION

The opioid supply is divided between prescription painkillers such as OxyContin, which end up on the black market, and illicit narcotics such as heroin and fentanyl. The two problems require different approaches.

Thanks to new practice guidelines and prescription surveillance programs, the number of high-dose legal opioid prescriptions dropped 41 percent between 2010 and 2016, and another 16 percent in 2017.[14] By the time the SUPPORT Act was signed, Congress had already passed the Prescription Drug Monitoring Act,[15] which included regulations regarding how many days' worth of opioids a person may be given at one time, as well as more sophisticated electronic databases to prevent customers from being written multiple prescriptions, stop diversion and forgery, and make it illegal for doctors to overprescribe. In a program called Operation Pilluted, the DEA proceeded to bust 280 providers for dispensing uncommonly large amounts of opioids.

While prescription opioids are often considered the "gateway" to stronger drugs, users actually enter the world of addictive drugs from various directions. In addition, overdoses are sometimes caused by the intentional use of multiple drugs or a drug that's been adulterated with other opioids. In particular, dealers have begun cutting cocaine, methamphetamines, crushed prescription opioids, and marijuana with fentanyl and other strong opioids to make them more powerful.

One of the more revealing by-products of the Prescription Drug

---

14   Singer, "The Administration's Fundamental Flaw on Opioid Addiction."
15   The PDMP Training and Technical Assistance Center, "Prescription Drug Monitoring FAQs."

Monitoring Act is that hospitals no longer include questions about pain management in their patient surveys. In the late 1990s, the VA (U.S. Department of Veterans Affairs) put into practice the concept of pain being the fifth "vital sign" based on the belief that pain went undertreated. The idea was subsequently adopted, in 2001, by the Centers for Medicare and Medicaid Services. Since this approach to pain has contributed to a rise in addiction and overdoses from prescription opioids, its effectiveness has been challenged and doctors are increasingly walking a fine line between satisfying regulators' (and the general public's) concern about addiction while keeping their patients as comfortable as possible.

Controlling the supply of illicitly manufactured drugs is even more difficult since the government cannot successfully find and prosecute every renegade chemist in America. While George Marquardt (see previous chapter) may have been an autodidact, chemist "visionary," and sociopath in more or less equal measure, it's likely that other, less eccentric chemists continue to brew up batches of equally deadly drugs. The best, if not only, way to deal with illegally manufactured drugs may not be by finding labs to raid, but by using the harm reduction strategies described below.

Ironically, the availability of powerful "homemade" opioids that can be manufactured without any natural source is making the issue of crop eradication far less relevant. There is less reason to implicate the poppy flower in the drug crisis. Nor is there any reason to develop complicated international political strategies to deal with its widespread cultivation. Nor is there any reason, in the long run, to spend valuable resources punishing other countries for manufacturing the drugs that are killing our citizens. Drug interdiction (and eradication) isn't just a zero-sum game, it's a negative-sum game. As we've seen, any supply reduction in one country will quickly be replaced by another. The "enemy" in the war on drugs is right here at home. As, ultimately, it always has been.[16]

## HARM REDUCTION

Addressing the opioid crisis involves reducing the individual harm caused by the drugs—which includes deadly infections from dirty needles as well as overdose deaths. Unfortunately, this is another area where stigma and prejudice prevent progress.

Making the overdose antidote naloxone more widely available has proven to be the most acceptable of these strategies. The drug works by blocking opioid receptors long enough to get the user to a hospital for further treatment. Since it's available as a nasal spray (Narcan), it is relatively easy to learn how to administer, so, in addition to being used by EMT personnel, it is becoming increasingly available at schools, libraries, and in the workplace (just like defibrillators).

Even those who don't think addicts "deserve" to be saved need to understand that naloxone can save the lives of first responders or friends of the overdose victim who come in contact with deadly amounts of fentanyl or carfentanil. Equally important is the fact that these drugs could be "weaponized" and used by terrorists to cause multiple deaths in a public place. In fact, an opioid was allegedly used by Russian police during a 2002 hostage crisis in a Moscow theater, which led to the deaths of the terrorists but also many of the moviegoers.

Another long-term harm-reduction strategy is to help overdose

---

16  Even China, which the United States has accused of flooding the market with opium, realizes that. See "China Says U.S. Should Do More to Cut Its 'Enormous' Opioid Demand."

survivors begin a MAT program (see section below) before discharge from an ER, by providing them with the first dose of methadone or Suboxone right there and creating a plan of services to support the user in the days ahead.

Needle-exchange programs are a far more controversial harm-reduction strategy, but they not only give providers and support personnel another opportunity to encourage users to enter treatment programs, they also prevent life-threatening contagious infections, such as hepatitis B and C and HIV, which, since they are sexually transmitted, can spread through the general public as easily as through a community of addicts.

"Batch" testing is another controversial but highly effective life-saving strategy. When a new drug appears on the street, users can anonymously submit samples to determine if it's what the dealer claims it is. In Portugal, users don't even have to go to special centers to have their drugs evaluated for purity because officials set up testing sites at popular gathering places such as music festivals and bars.

Heroin-Assisted Treatment (HAT) can be the last and best resort for users suffering from the most serious addictions—ones that don't respond to more conventional MATs. Currently available in parts of Switzerland, Germany, the Netherlands, and Canada, the treatment involves providing addicts with doses of the drug up to three times daily, often enabling them to function well enough to stay in school or keep their jobs.[17] HAT reduces both overdoses and serious infections (since the user is assured of getting the precise amount of an unadulterated drug injected safely). In addition, crime in neighborhoods with these clinics has gone down as much as 80 percent. Police

---

17   "Heroin-Assisted Treatment in Switzerland"; and Verthein et al., "Long-Term Effects of Heroin-Assisted Treatment in Germany."

departments in Germany that fought to keep the sites out now fight to be able to establish them in their communities.[18]

When hardened users are freed from the ravages of a lifestyle that's unavoidable if they have to buy opiates illegally to manage their addiction, they have the opportunity to rebuild their lives. Instead of the revolving door of jail or drug court (in either case being exposed to the criminal underground economy), they can get access to good medical and mental healthcare and free, clean heroin. They can regain their connections with family, establish a stable living situation and get (or keep) a job. In time, a number of these committed users seek help to convert from heroin to methadone or Suboxone or even to taper off at a pace of their choosing. By bringing their hidden opiate addictions out into the open, they can use heroin under the supervision of nonjudgmental healthcare providers and social workers, reducing the risk of overdose and death while, at the same time, creating the conditions for honest assessment of how their addiction is affecting their lives.[19]

Another controversial harm-reduction approach is the establishment of safe (or supervised) injection sites, where addicts can bring their own illegally-bought heroin to use at a place where naloxone is readily available in case the drug is contaminated. Some of these sites also offer batch testing and counseling.

In America, there is little support at the federal level for safe-injection sites, but a dozen cities and states including Seattle, Philadelphia, Ithaca, Baltimore, and Denver are seriously considering them. The mayor of New York, Bill de Blasio, has already expressed his support

---

18  Personal conversation, John Halpern with Torsten Passe. Also see Verthein et al., "Long-Term Effects of Heroin-Assisted Treatment in Germany."
19  Ibid.

for a pilot plan to open four safe injection sites, but officials are wary of opening them as long as federal authorities threaten sanctions and even arrest.[20] Ed Rendell, the former governor of Pennsylvania, believes the crisis is so severe that waiting for approval is not an option. He has dared the government to come arrest him for helping a nonprofit called Safehouse open a safe injection site in the city of Philadelphia, where 1,217 people died of overdoses last year.[21]

Another crucial harm-reduction strategy is the passage of Good Samaritan laws. In recognition of the vital seconds lost in emergencies (because people in the presence of someone overdosing fear repercussions if they report it) more than forty states and the District of Columbia have passed these laws. Good Samaritan laws provide various degrees of immunity from arrest, charge, or prosecution for anyone who calls 911 immediately if in the presence of someone who overdoses—even if the caller provided the drug and/or the needles.[22]

Objections to some or all of these treatments are based on the belief that drug use is a moral failing, abstinence is the only successful withdrawal strategy, and/or that the cost is prohibitive. Regardless, to reject proven solutions because of moral or cultural beliefs is not only counterproductive, it could be considered unethical—especially at a time when addiction often begins with the use of perfectly legal drugs. As far as cost, these strategies clearly yield long-term savings because of reductions in crime and ER use—not to mention the damage addiction does to families, businesses, and the community, as well as the positive impact of users returning to school or the workplace.

---

20  Azulay and Lewis, "De Blasio Endorses Creating a Safer Place for Drug Users."
21  Allyn, "'Come And Arrest Me.'"
22  National Conference of State Legislatures, "Drug Overdose Immunity and Good Samaritan Laws."

## PHARMACEUTICAL INNOVATION

Currently, there is no painkiller available—from aspirin and acetamino-phen to the stronger NSAIDs like high-dose ibuprofen and Toradol—that's as effective at relieving intense pain as the opioids, especially for treating certain cancers and end-of-life pain. The latest breakthroughs have involved the development of what are called COX-2 inhibitors, which affect an enzyme that causes pain and inflammation. There are also newer treatments for specific muscle or nerve pain (such as steroids, nerve blocks, and neurostimulators). Currently, there are ex-periments underway involving drugs that target other pain receptors, as well as some "nerve-growth factor drug inhibitors." In addition, now that cannabinoids are legal in some states, there is a rapidly growing body of promising anecdotal data from people using CBDs successfully to treat pain.[23]

The holy grail of pain treatment, however, is a less addictive opioid—a concept that has to raise the eyebrows, if not strike fear in the hearts, of anyone who's studied their opium history because, to date, that search has led only to more-addictive drugs—from morphine and heroin to "abuse-deterrent" OxyContin. (In fact, perhaps the FDA should issue a regulation outlawing the phrase "less addictive" in the promotion of any drug containing an opioid or opioid analog.) Regardless, there is some intriguing research into ways to activate opioid receptors without the risk of addiction.[24]

There are three main kinds of opioid receptors (referred to as *mu*, *kappa*, and *delta*) that are found in different parts of the nervous system.

---

23   Staines, "Pfizer/Lilly Non-Opioid Pain Drug Hits Mark in Phase 3 Trial."
24   Rosenblum et al., "Opioids and the Treatment of Chronic Pain." This report from the National Institutes of Health is a comprehensive overview of the science of opioids, historically and today.

Different opioids bind to them in different ways and have different side effects. An international team of scientists recently reported that they had identified the crystal structure of the kappa opioid receptor— which is considered less likely to be involved in addiction—and have developed an opioid that binds only with that one.[25]

Similarly, a biotech company in Canada reports it has isolated a specific gene in the opium poppy that encodes for thebaine (the alkaloid that is used in making oxycodone), which may eventually make it possible to use a technology called "microbial manufacturing" for commercial production of opioids from sugar—a process that the company hopes "will provide a basis from which to develop novel less-addictive opioids not currently accessible from the plant or traditional chemistries."[26] Researchers at Stanford are also working on a technology to produce less addictive opioids from sugar using a strain of yeast.[27]

Equally promising are further developments in the science of using a kind of protein known as a peptide. Originally extracted from frogs and sea mollusks, peptides are already used for treating a variety of conditions, and scientists are working on developing peptide medications that may be able to provide more targeted pain relief, without the risk of addiction.[28]

Since, as with any chronic disease, a person's vulnerability to addiction is based, in part, on his or her genetic makeup, scientists are also experimenting with ways to remove or deactivate "mutant" genes that trigger addiction—essentially creating a medicine for the

---

25  University of Southern California, "A Nonaddictive Opioid Painkiller."
26  Epimeron, "Discovery Opens Door for Synthetic Opioids with Less Addictive Qualities."
27  Stoye, "Biotech Breakthrough as Yeast Makes Painkillers from Sugar."
28  Lau and Dunn, "Development Trends for Peptide Therapeutics."

disease itself rather than one such as Suboxone that is designed to help with an "episode" of withdrawal and provide lasting prophylaxis against future relapse.[29]

## INSURANCE REFORM

Mental health diseases such as addiction are subject to coverage limitations that would never be considered when covering diseases of other organs—which, themselves, are often caused by addictive habits such as smoking, drinking, and overeating.[30] Even when the federal government or states have passed laws mandating parity between physical and mental illnesses, implementation is too often under-funded and poorly enforced. Medicaid itself has strict limitations on coverage for addiction treatments.

Some insurance companies are beginning to recognize that addiction is often comorbid with other physical health problems and that an integrated approach is less expensive over the long run. Others claim that, since consumers frequently switch insurance plans, it's difficult to factor in the costs of long-term addiction treatment—although this is done all the time for other chronic illnesses.

One solution is to have insurance plans follow the consumer himself or herself, which is, of course, controversial because it's considered a form of "universal healthcare." While covering addiction treatment as a preexisting condition may require more complex algorithms and more difficult policy debates, it's incumbent on the many politicians who believe that everyone deserves equal access to healthcare to find solutions.

---

29   "Gene Therapy for Addiction."
30   "The Kennedy Forum."

## MEDICATION ASSISTED TREATMENT (MAT)

Ironically, while opioids provide users with a simple solution to complex problems (physical pain, poverty, trauma, family history, depression, homelessness, and other psychiatric disorders); for society, developing effective treatments for addiction is a complex challenge for which there's no simple solution.

However, many experts believe that the combination of MAT with behavioral health services offers the best chance for success. MAT drugs include methadone, buprenorphine, Suboxone (buprenorphine + naloxone), and the opiate-antagonist naltrexone that also comes in a monthly injection (Vivitrol). For alcoholism, there are options like disulfiram (Antabuse) and acamprosate (Campral), and for nicotine dependence, there are nicotine replacement patches, gum, and lozenges as well as anti-craving varenicline (Chantix) and the antidepressant bupropion (Zyban/Wellbutrin), which can spoil the taste of tobacco.

Providers prescribe different replacement drugs based on the stage of a patient's addiction and the substance to which they are addicted. Methadone has been used in America since 1960 and helped thousands of people give up heroin. Like heroin, it is a "full agonist" (activator) of opioid receptors, but it has a much longer half-life, which makes it possible to achieve a steady-state—without the debilitating highs and lows of fast-acting heroin—so an addict can get back to work and everyday life in the same way as anyone else who takes medication for a chronic condition. Buprenorphine is a partial agonist that was introduced in the early 2000s to reduce or eliminate the cravings that usually accompany withdrawal. Suboxone, a combination of buprenorphine and the opioid blocker naloxone, is prescribed more often because, if taken as directed, it establishes an upper limit on the effect of street drugs. However, like any drug, Suboxone can be abused by snorting or injecting supplies that users with "take homes" sell on the black market. Since the user risks opiate withdrawal because of the naloxone component, however, Subutex (buprenorphine only) has more street value.

More promising for many patients is the recent development of extended-release injectable buprenorphine,[31] sold as Sublocade, and the newer Suboxone implants, which are sold as Probuphine. Both not only maximize compliance but, short of the user ripping out the implant, ensure the drug isn't abused.[32] Other manufacturers are exploring similar products. Regardless, the availability of month-long or longer medication is a crucial breakthrough in addressing opiate addiction: the merry-go-round of starting/stopping treatment and alternating treatment with relapse is reduced, along with the risk of these medications being diverted to the street. The treated patient also wakes up without the daily urgency to use because the long-acting treatment gives a steady-state dose of buprenorphine that doesn't increase/decrease day to day.

Unfortunately, there are many minor and major barriers that stop these drugs from being more widely or successfully administered, in particular the need of treatment centers to maintain financial stability in today's healthcare marketplace and spend valuable resources on capital campaigns to cover shortfalls. For example, some MAT treatments require less frequent monitoring and counseling, both of which are especially profitable. In addition, decisions on staffing and facilities can easily be made with an eye toward the bottom line, resulting in the overcrowding of patients or overworking of employees.

---

31  Bell, Belackova, and Lintzeris, "Supervised Injectable Opioid Treatment for the Management of Opioid Dependence."
32  Lazzara, "10 Ways to Get Off Opiates."

In an ideal world, patients would be evaluated by a doctor who could diagnose and treat their addictions along with all comorbid conditions without any regard for the economic consequences. However, there hasn't been the political will to deal with addiction holistically. In this vacuum, for-profit entities are entering the marketplace to provide care. While such centers are cropping up to address need, they, much like for-profit prisons, are providing a service that feeds off crisis. Politicians, policy makers, and policy implementers have yet to coalesce around a comprehensive public health strategy.

## *ALTERNATIVE* ADDICTION TREATMENTS

At a time when opioid addiction is a nationwide crisis, it is remarkable that there have been so many restrictions placed on drugs that might help address the crisis.

For decades, there have been clinical trials of powerful cancer drugs. Sometimes these chemicals prolong a patient's life. Sometimes they shorten it. Sometimes they have painful side effects. For example, many children who participated in multidrug chemotherapy trials during the 1950s not only didn't live longer, they endured horrific drug reactions. Yet, today, they are very rightly honored for the sacrifice they made so that other children could live.

Why aren't there more clinical trials for opioid addiction, which is potentially as lethal as any cancer? As it is, instead of going to a major medical center and enrolling in a clinical trial using alternative treatments, some addicts have resorted to going underground or to other countries to try experimental drugs.

Ibogaine, for example, a hallucinogen that comes from the root of a West African shrub, has been used to cure addiction, and a related drug known as 18MC may be able to take advantage of ibogaine's anti-addictive properties without the risk of hallucinations.[33] Ibogaine's use was first promoted by the late Howard Lotsof after he used it to cure his own heroin addiction in the 1960s, yet seventy years later, it remains illegal in most Western countries.[34] Ketamine, also known as a "party drug" called "Special K," has finally achieved limited FDA approval after proving it can bring immediate relief to some patients suffering from a major depressive episode and/or serious suicidal ideation, and may be able to start people on the road to recovery from addiction.[35]

Other addicts seeking treatment have turned to traditional rituals (in both indigenous and New Age settings) that involve a medicinal herb from Southeast Asia called *kratom*[36] and another using the Amazonian medicine *kambo*, also known as "Tree Frog Poison," which is rich in opioid peptides and has been shown to be helpful in managing withdrawal—at least in addicted rats.[37] Kratom is yet another natural drug that has ended up being tested on the street instead of in the lab—particularly because, like Special K, it can be used recreationally as well as therapeutically, depending on the dosage. However, according to a recent study, "human pharmacologic, pharmacokinetic, and clinical data are of low quality, precluding any firm conclusions regarding safety and efficacy." In other words, it's not that this promising drug has been proven to be ineffective, it's

33  Reiss, "Opioid Crisis."
34  Austin and Greenstein, "Ibogaine"; Alper, Lotsof, and Kaplan, "The Ibogaine Medical Subculture."
35  Lalanne et al., "Experience of the Use of Ketamine to Manage Opioid Withdrawal in an Addicted Woman."
36  Boyer et al., "Self-treatment of Opioid Withdrawal Using Kratom (Mitragynia Speciosa Korth)."
37  See "Opioid Peptides: Medicinal Chemistry, 69"; Daly, "How Amazonian Tree Frog Poison Became the Latest Treatment for Addiction."

that it hasn't received the attention that anecdotal reports indicate it deserves.[38]

Perhaps the most controversial alternative treatments are those that involve the hallucinogen LSD, which researchers had some success using in controlled medical settings in the 1970s before abandoning their experiments in the face of political pressure. One theory for LSD's effectiveness for addiction is that the rapid elevation in mood—usually lasting several weeks—makes it possible to endure the severe cravings during the early phases of recovery. This "afterglow" can also put the patient in a frame of mind that accelerates his or her ability and willingness to begin productive psychotherapeutic work to address the stress or trauma that lies under the addiction. Hallucinogens including DMT/ayahuasca, psilocybin, and mescaline have shown similar potential.[39]

The challenge, as with any drug under development, is to determine the therapeutic dosage that will deliver desired benefits without dangerous side effects. This has led some people to explore using "microdoses," which consist of just 5 percent or 10 percent of a traditional dose of one of the hallucinogens mentioned. Even so, all these drugs remain illegal to use without FDA approval because the agency believes them to have a high potential for abuse and dependency without established clinical protocols for medical safety.[40]

For years, even scientists have faced roadblocks in being permitted to use these drugs for research purposes. From the 1970s until the 1990s, LSD research was completely forbidden. Currently, however,

38  White, "Pharmacologic and Clinical Assessment of Kratom."
39  Halpern, "The Use of Hallucinogens in the Treatment of Addiction."
40  Chandler, "A Breakdown of Microdosing."

almost 600 researchers have now been approved to study these Schedule 1 drugs for a variety of conditions.[41]

Meanwhile valuable time has been wasted due to political pressure. People who potentially could have been helped by these drugs have died. It's a tragedy—and a situation that wouldn't be tolerated for experimental research on treatments for cancer or other conditions. Sometimes roadblocks are truly structural: the National Institutes of Health's National Institute on Drug Abuse is specifically prevented in its charter to fund any research into the therapeutic utility of drugs of abuse. The very government agency best familiar with the science of these drugs cannot explore how cannabis or LSD could help lessen drug addiction just because they are also drugs of abuse.

## BEHAVIORAL AND PSYCHOLOGICAL SUPPORTS

While medication can help patients deal with their physical cravings, the psychological nature of addiction means that replacement drugs are rarely a total solution. Effective treatment usually requires a comprehensive and tailored program that includes behavioral supports and psychological counseling.[42]

Different chronic opioid users require different supports depending on which "nature," "nurture," and environmental factors led to their addiction, as well as what experiences could trigger relapse. Many turn to Narcotics Anonymous, Alcoholics Anonymous, and other support groups to reinforce their sobriety. While there are conflicting studies about the overall success rates of AA and NA[43] (especially when

---

41   National Media Affairs Office, "DEA Speeds up Application Process."
42   Halpern, "Addiction Is a Disease."
43   Szalavitz, "After 75 Years of AA, It's Time to Admit We Have a Problem."

not supported by medication) for many patients, participation is an essential part of their ongoing sobriety. For those whose addiction is complicated by poverty and homelessness, community programs that provide food, shelter, training in life skills, education, job opportunities, and transportation can play a major role in recovery.

Some form of counseling is almost always helpful, since resolving psychological issues—particularly those related to trauma—can address the root causes of addiction. Cognitive behavioral training is one effective way to help people understand and avoid interactions and environments that trigger drug use. Another is to target basic lifestyle changes—such as diet and exercise—that can have the same positive effects as they do in treating any disease.

## INPATIENT TREATMENT

For many years, the government simply did not know what to do with debilitated addicts, so they relegated them to the same asylums that housed people with more "traditional" mental illnesses. After President Kennedy signed the Community Mental Health Act in 1963, these institutions began to be phased out and treatment was increasingly moved to community-based settings. Unfortunately, in many places there are now too few beds for those who need them, which leads to an increase in crime. For many years the largest inpatient facilities for drug possession in America were the Los Angeles County Jail and New York City's Rikers Island jail.[44] These are probably the worst possible places for long-term treatment—or, worse, trying to go cold turkey—since the very conditions promote an illegal drug subculture within and outside the prison.

---

44  Halpern, "Addiction Is a Disease."

In the end, an addict's wealth turns out to be a major factor not only in avoiding incarceration but also in the potential for successful inpatient treatment. Long after Medicare, Medicaid, and other traditional insurance coverages have run out, addicts with means can go to high-end facilities that offer individualized programs in comfortable residential settings where providers closely monitor their medication regimen and provide extensive psychological and behavioral treatments that include both the patient and his or her family. They also provide excellent diet, activities, life skills training, and a plan for a continuum of care after the patient returns to the community.

High-end facilities are expanding because the patients who can afford to go can be treated as long as medically necessary, rather than as long as their reimbursement continues to be authorized. Meanwhile, those with inadequate or no insurance are more likely to be released prematurely and, if they relapse, often end up being readmitted for care only after an expensive visit to the ER. The problem is caused largely because of the public perception that those needing treatment should be able to be cured of their addiction in the time frame prescribed by insurance companies (rather than their doctors). For-profit programs are gaining traction at huge, unnecessary costs because we lack a unified program of treatment at the national level that would also welcome piloting innovation. Without national support that accepts the opioid crisis as a national disaster, policy still slides in the direction of placing morality ahead of protection of life.

### EDUCATION

The most cost-effective and simplest approach to treating addiction is, of course, to prevent it in the first place, through a variety of prevention strategies, most importantly education.

In the 1980s, President Reagan and his wife Nancy led an "educational" campaign to persuade kids to "just say no," and LAPD Chief Daryl Gates founded the school-based DARE program (Drug Abuse Resistance Education). Unfortunately, many adolescents are drawn to take risks during adolescence, and to suggest that we can "scare" them into not taking drugs is a contradiction in terms. One drug policy adviser to presidents George W. Bush and Barack Obama described these failures bluntly, saying that "while these, and other similar programs, cost billions of dollars and generated a lot of publicity, the general consensus is that they either had no effect or in some cases maybe even a perverse effect that some of the kids who saw the most ads actually said they were more likely to try marijuana rather than less."[45] But ONDCP-funded ads like "this is your brain on drugs" proved quite popular with their parents.

As far as using suspension or expulsion to "educate," it's essential to repeat yet again that addiction is a disease. It is a chronic disease. We don't suspend kids for having diabetes or asthma or multiple sclerosis or other chronic diseases. Instead, we provide them with the supports and accommodations they require so they can stay in school. Students who are at risk for addiction or who have become addicted deserve exactly the same level of services.

The education, prevention, and treatment programs in schools and community youth organizations with the best outcomes do not just educate students about the risks of drugs but teach them about brain health holistically, and address their individual questions and concerns. Discussions take the form of age-appropriate explanations of how the brain works, showing the impact on the developing brain of

---

45   Keith Humphreys, professor of psychiatry at Stanford University. Quoted in Shapiro, "A Look at the Effectiveness of Anti-Drug Ad Campaigns."

not only drugs, alcohol, and tobacco (or excessive screen time, which is currently the most common youth addiction), but also how diet, exercise, and other lifestyle choices can heighten brain function. This information need not be confined to a specific course such as health or biology but incorporated throughout the curriculum—in courses such as history, sociology, English, and others. Rather than simply make substance-abuse prevention a "top-down" discipline that for many students offers a road map for rebellion, holistic brain education empowers students to take more responsibility for their own decisions. Students inherently know when they are being ask to believe propaganda that's backed without any independent thought: education cannot just be about harms and dangers without acknowledging what these drugs offer that make them attractive to use in the first place.

Unfortunately, while most reports about ways to reduce opioid abuse insist on the need for these "evidence-based" initiatives, there's a good bit of disagreement about how that evidence is measured. Moreover, teachers are often not given adequate training in age-appropriate substance-abuse education or not given the time to provide that education effectively.

The Substance Abuse and Mental Health Services Administration has an online Evidence-Based Practices Resource Center that provides links to more than 400 programs that cover drugs (including specific ones for marijuana and opioids), as well as alcohol, tobacco, suicide, and other topics that can be implemented in schools. Unfortunately, it's hard for an educator to find the right program because within any grade or group, there are significant differences in developmental age, culture, economic background, family structure, et cetera.[46]

---

46   "EBP Resource Center."

Ideally, schools would screen students for risk of (or existing) substance-use disorders (SUDs), just as they do for other factors that could significantly impact the child's ability to learn—such as hearing, vision, dental health, and obesity. By doing so, students could be given the services they need *before* they have manifested a serious drug or alcohol problem—for example, addressing family dynamics or psychiatric problems. It would be naïve to think that interventions like this would totally prevent a student from developing an SUD, but it could go a long way to preparing both the child, his or her peers, and the school to more quickly and successfully recognize it and respond.

Another common feature in prevention and harm-reduction programs involves engaging students in sports and extracurricular activities in which substance abuse can affect performance or disqualify the student altogether. Even here, while there may be a place for random drug testing, the efforts need to focus on prevention and treatment. Rather than a policy of three-strikes-and-out, schools could consider a one-strike-and-in-treatment approach.

Finally, there is the "social norms" strategy that attempts to prove to students that use and abuse of drugs is less common than they think, thereby limiting the affect of any real or imagined social pressure. While the social-norms approach can be counterproductive for at-risk students whose drug use is predicated on experimentation, it has proven effective among populations that are not predisposed to drug use or those that need support in recovery. One example at the college level is the option of living in "sober" dorms, in which drugs and alcohol are not allowed and counseling is more readily available. With millions of college-age young adults dealing with substance abuse, this kind of housing offers a home where they can continue their recovery in a supportive environment. Some are even beginning to offer medication-assisted treatment.[47]

Ultimately, however, in terms of school and community solutions to young-adult substance abuse, it's important to remember that all a government program, caring teacher, sensitive school psychologist, or even a parent can do is provide the information students need to make their decisions (whether "right" or "wrong"), model behavior that's empowering instead of debilitating, provide unconditional and nonjudgmental support regardless, and hope for the best.

\* \* \*

As the United States begins its third decade of its worst health epidemic since HIV/AIDS—with more than 70,000 Americans dying per year from overdoses and thousands more brought back from the brink in ambulances and emergency rooms, the crisis can seem not just overwhelming but utterly unsolvable.

History has a lot to teach us about this crisis, but if we're going to learn it, we need to let go of the conceit of modernity: the notion, in other words, that this crisis is worse, or fundamentally different, than any that has come before. We have to realize once and for all how absurd it is to think harsh penalties will reduce drug use when we've seen how those same harsh penalties have failed time after time; how preposterous it is to imagine that eradicating one source of supply will solve anything when we've seen again and again how quickly a new one will arise to take its place; how arrogant it is to think we can legislate behavior when governments have failed for centuries to do just that.

---

47  Kaplan, "A Growing Number of Sober Programs Support College Students Recovering from Addiction."

Most of all, we have to confront the unreasonable fear and unenlightened self-interest that enables our leaders to argue that addiction is a choice to be punished rather than an illness to be cured. In addition, we have to resist the seduction of seemingly easy solutions such as guarding borders, locking up users, or telling kids to just say no.

Perhaps the hardest, most crucial step is to admit the crisis will never end. As long as one person dies from an overdose, it will be a crisis for his or her family, friends, and community. Let's not have that person die due to our fear and ignorance. Let's show that we have the wisdom, guts, humility, and compassion to save the lives of thousands of others.

# Afterword

I think about my friend Paul Roderick every day—as well as patients who have died equally tragic deaths.

And yet I remain infectiously optimistic. Because I believe that every time you save one person you save the world.

Time and again, I've seen people survive the deadly grip of opioids. Even if Paul wasn't one of them, I hope his story, too, can save one person, maybe more, and thereby save the world.

It's the height of arrogance to think we know the significance of another's life—that we understand the challenges they face and are in a position to judge what they do in response. Miles Davis and Stan Getz survived their addictions. Charlie Parker and Billie Holiday didn't. Neither did Jimi Hendrix, Janis Joplin, Prince, or Tom Petty. All of them died from accidental or intentional overdoses directly related to the excruciating pain—emotional and physical—they endured while creating music that inspires us to this day. Who are we to condemn their choices?

We can't save everyone, but it's time to stop playing the moral judgment game and focus on helping people who ask us for help. It's time to think about not just reducing harm but maximizing health, as you can't have one without the other.

It's also time to stop overloading the criminal justice system with people whose only crime is suffering from a particularly debilitating disease. It's not a question of being tough or soft on crime...it's about being *smart* on crime.

Let's face it, we humans do stupid stuff. We always have and we always will. The general public has, at last, begun to acknowledge that drug users and abusers aren't "those people." They are us— our children, our siblings, our parents, our extended families, and our friends.

Thankfully, we don't live in a totalitarian state. Unlike China, we don't send users to reeducation camps and give the death penalty to dealers—with only the mere shadow of a trial. Unlike Singapore or the Philippines, we don't execute people caught possessing even minimal amounts of drugs for personal use.

We are in the midst of a public health emergency. Our patients don't have time for their medical professionals to be intimidated by politicians' opinions about "acceptable" ways to treat addiction, or their endless and disingenuous arguments about how to fund that treatment. No, our patients need us to focus on doing everything we can to help them, regardless of the significant scientific, philosophical, and personal challenges we face in the process. Not doing so is equivalent to malpractice: doing nothing, staying silent is being negligent to our obligation to do some good.

Several years ago, I was teaching a class at Harvard Medical School and asked my students whether they wanted to be doctors or healers. Some chose one and some the other. I told them they were all wrong. We need to be both. Doctor and healer. Empiricist and empath.

In the years to come, research will empower healthcare professionals with radically more potent tools against drug addiction, but, that said, our most potent treatment will always be in our hearts.

Paul reminds me of that every day.

——*John H. Halpern, MD*

# Acknowledgments

First, a special thank you to Rick Doblin, PhD, president of the Multidisciplinary Association for Psychedelic Studies, for initiating a wonderful collaboration and friendship by introducing us to each other. Thanks to our agent Johanna Maaghul for her boundless enthusiasm, encouragement, and tenacious advocacy on behalf of two rather idiosyncratic writers. At Hachette, our editor David Lamb recognized the importance of the project, kept us focused, and made sure our message was perfectly clear. Copy editor Bill Warhop did a remarkable job of identifying inconsistencies and ambiguities while dealing with several thousand years of strange names; and production editor Cisca Schreefel made the whole thing come together. In addition, our publicist, Joanna Pinsker, and marketer, Quinn Fariel, helped this important story get the attention it needs to make a difference. We'd also like to thank Mandy Kain for the book's provocative cover design and Thomas Louie for its interior design.

Professor Nico Cellinese and Egyptologist Paula Veiga clarified some of the latest archaeological and botanical discoveries (although any questionable speculations are totally our responsibility). Solomon Polshek and Jamie Walker provided some critical research early on. In terms of research, we are most indebted to Susan Shumaker, especially for uncovering facts and reliable sources for the convolutions of American drug policy in the twentieth and twenty-first centuries; Susan also took on the byzantine job of obtaining images and rights. Thanks to Than Saffel for creating maps that made it clear how opiates made it around the Mediterranean in ancient times, and up the Pearl River more recently.

**John Halpern, MD:** My thanks and appreciation to Marilyn Halpern, PhD; Emily Halpern Lewis; T. Mark Halper, J.D., for their many helpful suggestions for this book. Thanks to Andrea Sherwood, PhD and Jeanne Bauer for their support and encouragement. Thank you to my many patients over the years who have validated my philosophy that care really is central to effective clinical practice. Those that didn't survive their addiction are never forgotten. Thank you to many mentors, colleagues, and friends over the years who agree that we must not dream those dreams but live those dreams: Harrison G. Pope Jr., MD; Torsten Passie, MD, MA; Jack Mendelson, MD; Katherine Bonson, PhD; Steven Grant, PhD; Neal Goldsmith, PhD; David Blinder, PhD; Ari Mello; Gunnar Stefan Wathne; Donald Barr; the Hon. Victor J. Clyde and family, Eugene Spirit Eagle Beyale, Roy Haber, and Alfred Savinelli. Finally, and most important, thank you to my darling Ann LeBlanc and my son Noah—a happy home makes for a happy doctor who from time to time can offer then something even more.

**David Blistein:** Even in the small town of Brattleboro, Vermont, substance abuse is a critical issue. So, I'm grateful for the opportunities the Windham County Guardian ad Litem program has given me to see firsthand how drugs and alcohol affect children and families. I'd also like to thank Amy's Bakery, Echo Lounge, and Duo Restaurant in Brattleboro for providing inspiring "office space" throughout the writing of this book; as well as the many friends who listened patiently to my latest fulminations about the agony and ecstasy of the most powerful flower on earth. Also, a special thanks to Professor Adeline Hofer, who heard most all of it first.

Over the last five decades I've talked, brainstormed, and contemplated the fundamental moral and mind-altering issues underlying this book with countless people, but I want particularly to thank Virginia and all my Montana friends, as well as Ken Burns, Joe Kohout, Joe Marks, and Tom Yahn.

Finally, my wife, Wendy O'Connell, and I have now spent forty-two "some-odd" years exploring the mysteries of mind, heart, and spirit. Happily, it keeps getting more interesting.

# Bibliography

ACLU. "Just Say No to Random Drug Testing: A Guide for Students." American Civil Liberties Union, 2018. https://www.aclu.org/other/just-say-no-random-drug-testing-guide-students.

A.D.A.M. Consumer Health. "Sexually Transmitted Diseases." Penn State Medical Center, March 24, 2015. http://pennstatehershey.adam.com/content.aspx?productId=107&pid=33&gid=000151.

"Addiction A-Z: Uniform State Narcotic Drug Act." Addiction.com. Accessed July 4, 2018. https://www.addiction.com/a-z/uniform-state-narcotic-drug-act/.

"Addiction Deathmatch: Heroin vs. Sex." Blvd Treatment Centers, April 29, 2016. https://www.blvdcenters.org/blog/addiction-deathmatch-heroin-vs-sex.

Africa, Thomas W. "The Opium Addiction of Marcus Aurelius." *Journal of the History of Ideas* 22, no. 1 (March 1961): 97–102. doi.10.2307/2707876.

Aggrawal, Anil. "The Story of Opium." In *Narcotic Drugs*, Chapter 2. New Delhi: National Book Trust, 1995.

Agus, David. *The End of Illness*. New York: Free Press, 2011.

Albuquerque, Afonso de. *The Commentaries of the Great Afonso de Albuquerque, Second Viceroy of India*. Vol. 3. Translated by Walter de Gray Birch. London: Hakluyt Society, 1774. Google Books.

Alkhateeb, Firas. "Al-Zahrawi—The Pioneer of Modern Surgery." *Lost Islamic History* (blog), December 16, 2012. http://lostislamichistory.com/al-zahrawi/.

Allyn, Bobby. "'Come And Arrest Me': Former Pa. Governor Defies Justice Department On Safe Injection." NPR.org, October 10, 2018. https://www.npr.org/sections/health-shots/2018/10/10/656268815/come-and-arrest-me-former-pa-governor-defies-justice-department-on-safe-injection.

Alper, K. R., H. S. Lotsof, and C. D. Kaplan. "The Ibogaine Medical Subculture." *J. Ethnopharmacology*, August 21, 2007.

Altonen, Brian. "Opium Experiments by a Quaker." *Brian Altonen, MPH, MS* (blog), January 1, 2010. https://brianaltonenmph.com/6-history-of-medicine-and-pharmacy/hudson-valley-medical-history/1795-1815-biographies/quaker-shadrach-ricketson-md/the-opium-experiments/.

Altonen, Brian, MPH, MS. "Shadrach Ricketson, Quaker MD." *Public Health, Medicine and History* (blog), December 26, 2009. https://brianaltonenmph.com/6-history-of-medicine-and-pharmacy/hudson-valley-medical-history/1795-1815-biographies/quaker-shadrach-ricketson-md/.

Amkreutz, Luc, Fabian Haack, Daniela Hoffman, and Ivo van Wijk. *Something Out of the Ordinary? Interpreting Diversity in the Early Neolithic Linearbandkermaik and Beyond*. Newcastle upon Tyne: Cambridge Scholars Publishing, 2016.

Anderson, John Lee. "The Man in the Palace." *The New Yorker*, June 6, 2005.

Andrews, Edward Deming, and Faith Andrews. *Work and Worship Among the Shakers*. 1974. Reprint, New York: Dover, 1982.

Andrews, Evan. "Nine Things You May Not Know About the Ancient Sumerians." History.com, December 16, 2015. http://www.history.com/news/history-lists/9-things-you-may-not-know-about-the-ancient-sumerians.

Andrews, Stefan. "Heroin, Prescribed for Coughs and Headaches, Was a Trademarked Medicine Produced by Bayer Company." *The Vintage News* (blog), November 26, 2017. https://www.thevintagenews.com/2017/11/26/coughs-and-headaches/.

"Angell, China and Opium: University of Michigan Heritage." Accessed December 19, 2018. https://heritage.umich.edu/stories/angell-china-and-opium/.

Ansley, Norman. "International Efforts to Control Narcotics." *Journal of Criminal Law, Criminology, and Police Science* 50, no. 2 (July 1959): 105. doi.10.2307/1140682.

Anthony, Robert J., ed. *Elusive Pirates, Pervasive Smugglers: Violence and Clandestine Trade in the Greater China Seas*. Hong Kong: Hong Kong University Press, 2010.

Apostolakis, Efstratios, Nikolaos A. Papakonstantinou, Nikolaos G. Baikoussis, and Georgia Apostolaki. "Alexander the Great's Life-Threatening Thoracic Trauma." *Korean Journal of Thoracic and Cardiovascular Surgery* 51, no. 4 (August 2018): 241–246. doi.10.5090/kjtcs.2018.51.4.241.

Armstrong, David. "Sackler Embraced Plan to Conceal OxyContin's Strength from Doctors, Sealed Testimony Shows." *ProPublica*, February 21, 2019. https://www.propublica.org/article/richard-sackler-oxycontin-oxycodone-strength-conceal-from-doctors-sealed-testimony.

Arnold, Willis R. "Setting Precedent, a Federal Court Rules Jail Must Give Inmate Addiction Treatment." NPR.org. https://www.npr.org/sections/health-shots/2019/05/04/719805278/setting-precedent-a-federal-court-rules-jail-must-give-inmate-addiction-treatmen.

Askwith, Richard. "Heroin, Bayer and Heinrich Dreser." *Sunday Times*, September 13, 1998, opioids.com edition. https://www.opioids.com/heroin/heroinhistory.html.

Astyrakaki, Elisabeth, Alexandra Papaioannou, and Helen Askitopoulou. "References to Anesthesia, Pain, and Analgesia in the Hippocratic Collection." *Anesthesia & Analgesia* 110, no. 1 (January 2010): 188. doi.10.1213/ane.0b013e3181b188c2.

Athenaeum Centenary. *The Influence and History of the Boston Athenaeum from 1807–1907*. Boston: Boston Athenaeum, 1907.

Aurelius, Marcus. *Meditations*. Translated by George Long. Classic Club. Roslyn, NYC: Walter J. Black, 1945.

Austin, Gregory A. *Perspectives on the History of Psychoactive Substance Use*. Rockville, MD: National Institute on Drug Abuse, 1978.

Austin, Paul, and Louis Greenstein. "Ibogaine." *The Third Wave*, 2018. https://thethirdwave.co/.

Azulay, Zoe, and Danny Lewis. "De Blasio Endorses Creating a Safer Place for Drug Users." *WNYC News*. New York: NPR, May 3, 2018. https://www.wnyc.org/story/creating-safer-place-drug-users/.

Bacon, Francis. *Historie Naturall and Experimentall, of Life and Death. Or of the Prolongation of Life*. Edited by William Rawley. Early History of Medicine, Health & Disease. William Lee and Humphrey Mosley, 1638. http://name.umdl.umich.edu/A01454.0001.001.

Bacon, Sir Francis. "Francis Bacon on Opium, Coffee and Tobacco, Etc." *Faust*, April 20, 2016. https://www.faust.com/francis-bacon-on-opium-coffee-and-tobacco-etc/.

Bahn, So Yeong, Byung Hoon Jo, Yoo Seong Choi, and Hyung Joon Cha. "Control of Nacre Biomineralization by Pif80 in Pearl Oyster." *Science Advances* 3, no. 8 (August 2, 2017). doi.10.1126/sciadv.1700765.

Barbier, André. "The Extraction of Opium Alkaloids." United Nations Office on Drugs & Crime, 1950. http://www.unodc.org/unodc/en/data-and-analysis/bulletin/bulletin_1950-01-01_3_page004.html.

Barbosa, Duarte, Mansel Longworth Dames, and Fernão de Magalhães. *The Book of Duarte Barbosa; an Account of the Countries Bordering on the Indian Ocean and Their Inhabitants*. Works Issued by the Hakluyt Society, 2d Ser., No. 44, 49. London: Printed for the Hakluyt Society, 1918. HathiTrust.

Barter, James. *Opium*. Drug Education Library. Farmington Hills, MI: Lucent Books, 2005.

Barton, William P. C. *Outlines of Lectures on Materia Medica and Botany*. Vol. 2. Philadelphia: Joseph G. Auner, Bookseller, 1827. Google Books.

"Battle of Swally Hole." Encyclopedia Britannica. Accessed April 2, 2018. https://www.britannica.com/topic/Battle-of-Swally-Hole.

Beach House Center for Recovery. "Addiction in Medical Professionals." AddictionCenter. Accessed September 15, 2018. https://www.addictioncenter.com/addiction/medical-professionals/.

Bebinger, Martha. "How Profits from Opium Shaped 19th-Century Boston." *CommonHealth*. Boston, July 31, 2017. http://www.wbur.org/commonhealth/2017/07/31/opium-boston-history.

Bell, James, Vendula Belackova, and Nicholas Lintzeris. "Supervised Injectable Opioid Treatment for the Management of Opioid Dependence." *Drugs* 78 (August 21, 2018). doi.10.1007/s40265-018-0962-y.

Bergreen, Laurence. *Marco Polo: From Venice to Xanadu*. New York: Alfred A. Knopf, 2007.

Bernath, Jeno, ed. *Poppy: The Genus Papaver*. Taylor & Francis e-Library, 2005. Google Books.

"Billie Holiday: African American Singer." Black History in America. Accessed July 10, 2018. http://www.myblackhistory.net/Billie_Holiday.htm.

Blakeslee, George H. *China and the Far East*. Clark University Lectures. New York: Thomas Y. Crowell, 1910.

Blue, Lucy. "Late Bronze Age Trade in the Eastern Mediterranean—Shipwrecks and Submerged Worlds—University of Southampton." *FutureLearn*, 2017. https://www.futurelearn.com/courses/shipwrecks/0/steps/7966.

Bollinger, Lee C., and Ursula M. Burns. "Ending the Opioid Crisis: A Practical Guide for State Policymakers." *Center on Addiction*, October 2017.

Bonczar, Thomas P. "Prevalence of Imprisonment in the U.S. Population, 1974–2001." *Bureau of Justice Statistics*, 2013, 12.

Bonnie, R. J., et al. *Criminal Law*. 2nd ed. New York: Foundation Press, 2004.

Booth, Martin. *Opium: A History*. London: Simon & Schuster, 1996.

Bostrum, Nick, and David Pearce. "Opium Timeline." Accessed March 29, 2018. https://www.opioids.com/timeline/index.html.

Boyer, Edward W., Kavita M. Babu, Jessica E. Adkins, Christopher R.

McCurdy, and John H. Halpern. "Self-Treatment of Opioid Withdrawal Using Kratom (Mitragynia Speciosa Korth)." *Addiction* 103, no. 6 (June 1, 2008): 1048–1050. doi.10.1111/j.1360-0443.2008.02209.x.

Braswell, Sean. *From Drug Deals to the New Deal.* OZY. Accessed January 12, 2018. http://www.ozy.com/flashback/the-drug-that-bankrolled-some-of-americas-great-dynasties/40555.

Brecher, Edward M. "Opium Smoking Is Outlawed." *Consumers Union Report on Licit and Illicit Drugs,* 1972. http://druglibrary.org/schaffer/library/studies/cu/cu6.htm.

Bridgman, Elijah. *The Chinese Repository VIII.* Vol. 8. Canton: Elijah Bridgman, 1838.

————. *The Chinese Repository XI.* Vol. 11. Canton: Elijah Bridgman, 1842. HathiTrust.

"A Brief History of Heroin." Heroin.net. Accessed May 26, 2018. https://heroin.net/about/a-brief-history-of-heroin/#discovery.

Brody, Richard. "Judy Garland's Hollywood Unravelling, Through the Eyes of Her Husband and Producer Sid Luft," *New Yorker,* February 25, 2017. https://www.newyorker.com/culture/richard-brody/judy-garlands-hollywood-unravelling-through-the-eyes-of-her-husband-and-producer-sid-luft.

Brogan, Olwen. "Trade Between the Roman Empire and the Free Germans." *Journal of Roman Studies* 26 (1936): 195–222. doi.10.2307/296866.

Brown, Kate. "'This Is Reprehensible': Nan Goldin Responds to News That Richard Sackler Has Patented an OxyContin-Addiction Drug." *Artnet,* September 11, 2018. https://news.artnet.com/art-world/nan-goldin-sackler-statement-purdue-1346279.

Browne, Sir Thomas. *The Works of Sir Thomas Browne.* Vol. 3. London: Henry G. Bohn, 1852. Google Books.

Brownfield, William R. "Briefing on the International Narcotics Control Strategy Report." Special Briefing. US Department of State, March 2018. http://www.state.gov/j/inl/rls/rm/2017/268146.htm.

Brownstein, M. J. "A Brief History of Opiates, Opioid Peptides, and Opioid Receptors." *Proceedings of the National Academy of Sciences of the United States of America* 90, no. 12 (June 15, 1993): 5391–5393.

Bryan, Cyril P. "Ebers Papyrus." In *The Papyrus Ebers/Translated from the German Version,* translated by H. Joachim. London, 1930. Reprint, Chicago: Ares Publishers, 1974.

Buchanan, Rita. "A Peek into History: Shaker Herbs." *Mother Earth Living,* March 1997.

———. *The Shaker Herb and Garden Book*. Boston: Houghton Mifflin, 1996.

"Bulletin on Narcotics." New York: United Nations Office on Drugs & Crime, 1957. https://www.unodc.org/unodc/en/data-and-analysis/bulletin/bulletin_1957-01-01_2_page007.html.

Burke, Edmund. "Edmund Burke Quotations." In *Prose Quotations from Socrates to Macaulay*, edited by S. Austin Allibone. Philadelphia: J. B. Lippincott, 1880. http://www.bartleby.com/349/authors/33.html.

Burroughs, William S. "God's Own Medicine." In *The Adding Machine: Selected Essays*, New York: Grove Press, 1985. Kindle edition.

Buzzkill, Professor. "Marco Polo." *Professor Buzzkill—History's Myths Debunked* (blog), October 18, 2016. http://professorbuzzkill.com/marco-polo/.

Callery, M. M., and Yvan. "Emperor Tao-Kwang and the Opium War." *The Open Court* 1901, no. 9, Article 6 (1901): 7.

Campsie, Alison. "The Victorian Scots Who Brought Opium to China." *The Scotsman*, July 25, 2016. https://www.scotsman.com/lifestyle/the-victorian-scots-who-brought-opium-to-china-1-4186123.

Cantor, Donald J. "The Criminal Law and the Narcotics Problem." *Journal of Criminal Law, Criminology, and Police Science* 51, no. 5 (January 1961): 512. doi.10.2307/1141414.

Carod-Artal, F. J. "Psychoactive Plants in Ancient Greece." *Neurosciences and History* 1, no. 1 (2013): 28–38.

Carrero, Edgar. "Militarization Is No Answer to the Failures of the Drug War in Latin America." *Harvard College Law Review*, January 19, 2018.

Carter, A. J. "Dwale: An Anaesthetic from Old England." *British Medical Journal* 319, no. 7225 (December 18, 1999): 1623–1626. doi.10.1136/bmj.319.7225.1623.

Carter, Jimmy. "Drug Abuse Message to the Congress." August 2, 1977. http://www.presidency.ucsb.edu/ws/?pid=7908.

Cartwright, Mark. "Eunuchs in Ancient China." In *Ancient History Encyclopedia*, July 27, 2017. https://www.ancient.eu/article/1109/eunuchs-in-ancient-china/.

Castro, Joaquim Magalhães de. "Gaspar Da Cruz, the Dominican Traveller." *Oclarim*, June 3, 2016. http://www.oclarim.com.mo/en/2016/06/03/gaspar-da-cruz-the-dominican-traveller-1/.

Celsus, Aulus Cornelius. *De Medicina*. Edited by Bill Thayer. Loeb Classic Library-Reproduction. Harvard University Press, 1935. http://penelope.uchicago.edu/Thayer/E/Roman/Texts/Celsus/3*.html.

———. "De Medicina." Accessed January 7, 2018. http://www.perseus.tufts.edu/hopper/text?doc=Perseus:text:1999.02.0142.

Chandler, Carmen R. H. "A Breakdown of Microdosing." *Healthline,* 2018. https://www.healthline.com/health/beginners-guide-to-microdosing#13.

Chaplan, Michael. *The Urban Treasure Hunter: A Practical Handbook for Beginners.* Square One Publishers, 2005.

Chapman, Michael. "Pragmatic, Ad Hoc Foreign-Policy Making of the Early Republic: Thomas H. Perins's Boston-Smyrna-Canton Opium Model and Congressional Rejection of Aid for Greek Independence." *International History Review* 35, no. 3 (2013): 449–464.

Chapman, Michael E. "Taking Business to the Tiger's Gate: Thomas Handasyd Perkins and the Boston-Smyrna-Canton Opium Trade of the Early Republic." *Journal of the Royal Asiatic Society Hong Kong Branch* 52 (2012): 7–28.

"Charles Gabriel Pravaz." *Canadian Medical Association Journal* 13, no. 11 (November 1923): 839–840.

Charles, J. B. "Criminal Justice Reformers Are Hooked on Drug Courts. They Should Kick the Habit." The Hill, August 5, 2017. https://thehill.com/blogs/pundits-blog/crime/345371-criminal-justice-reformers-are-hooked-on-drug-courts-they-should.

Chavkin, Wendy. "Cocaine and Pregnancy—Time to Look at the Evidence." *JAMA* 285, no. 12 (March 28, 2001): 1626–1628. doi.10.1001/jama.285.12.1626.

Chen, David W. "Chinatown's Fujianese Get a Statue." *New York Times,* November 20, 1997, sec. N.Y. / Region.

Chouvy, Pierre Arnaud. *Opium: Uncovering the Politics of the Poppy.* Cambridge, MA: Harvard University Press, 2010.

"A Chronicle of Federal Drug Law Enforcement: The Early Years." *Drug Enforcement,* February 1979.

Clark, Gerald. *Get Happy: The Life of Judy Garland.* New York: Random House, 2000.

"Classics of Traditional Chinese Medicine: Emperors and Physicians." Exhibitions, October 19, 1999. https://www.nlm.nih.gov/exhibition/chinesemedicine/emperors.html.

Cobb, Cathy, and Harold Goldwhite. *Creations of Fire: Chemistry's Lively History.* Cambridge, MA: Perseus, 1995.

"Cocaine and Opium Production Worldwide Hit 'Absolute Record Highs'— Major Threat to Public Health Says UN Study." New York: United Nations Office on Drugs & Crime, June 26, 2018. https://news.un.org/en/story/2018/06/1013072.

Code of Federal Regulations. "Temporary Scheduling Order: Temporary

Placement of Fentanyl-Related Substances in Schedule I." Rules—2018. Washington, DC: Department of Justice—Drug Enforcement Administration, February 6, 2018. https://www.deadiversion.usdoj.gov/fed_regs/rules/2018/fr0206_4.htm.

Coleridge, Samuel Taylor, Ernest Hartley Coleridge, Cynthia Morgan St. John, and Wordsworth Collection. *Letters of Samuel Taylor Coleridge.* Boston: Houghton, Mifflin and & Co/Riverside, 1895. Internet Archive.

Collard, David. "Altered States of Consciousness and Ritual in Late Bronze Age Cyprus." PhD thesis, University of Nottingham, 2011. http://eprints.nottingham.ac.uk/12322/.

*Colliers.* "A San Francisco Public Bonfire That Cost $20,000." March 7, 1914.

Columbus, Christopher, and Paolo del Pozzo Toscanelli. *The Journal of Christopher Columbus (during His First Voyage, 1492–93) and Documents Relating the Voyages of John Cabot and Gaspar Corte Real.* Translated by Sir Clements R. Markham. London: Hakluyt Society, 1893. Internet Archive.

Commissioner of Internal Revenue. "Annual Reports 1914–1916." Washington, DC: Treasury Department. HathiTrust.

Comptroller General. "Gains Made in Controlling Illegal Drugs, Yet the Drug Trade Flourishes." Report to Congress. Washington, DC: Government Accounting Office, October 25, 1979.

Constable, Pamela. "Opium Use Booms in Afghanistan, Creating a 'Silent Tsunami' of Addicted Women." *Washington Post*, June 19, 2017, sec. Asia & Pacific.

"The Contras, Cocaine, and U.S. Covert Operations: National Security Archive Electronic Briefing Book No 2." National Security Archive. Accessed August 21, 2018. https://nsarchive2.gwu.edu//NSAEBB/NSAEBB2/index.html.

Cottle, Joseph. *Reminiscences of Samuel Taylor Coleridge and Robert Southey.* 2nd ed. London: Houlston and Stoneman, 1848. Google Books.

Courtwright, David T. "A Century of American Narcotic Policy." In *Treating Drug Problems: Commissioned Papers on Historical, Institutional, and Economic Contexts of Drug Treatment.* Vol. 2. Washington, DC: National Academies Press, 1992.

———. *Dark Paradise: Opiate Addiction in America before 1940.* Enlarged. Cambridge, MA: Harvard University Press, 2001.

Coyne, Christopher J., and Abigail R. Hall. "Four Decades and Counting: The Continued Failure of the War on Drugs." Cato Institute, April 12, 2017. https://www.cato.org/publications/policy-analysis/four-decades-counting-continued-failure-war-drugs.

Crafts, Wilbur Fisk, Sarah Jane Crafts, Mary Leitch, and Margaret W. Leitch. *Intoxicants & Opium in All Lands and Times, a Twentieth Century Survey of Intemperance*. Washington, DC: Reform Bureau, 1900.

CRC Health Group. "What Is LAAM? How Is It Different from Methadone?" *CRC Health* (blog). Accessed October 1, 2018. http://www.crchealth.com/ addiction/heroin-addiction-treatment/heroin-detox/laam/.

Cronkite, Susan-Marie. "The Sanctuary of Demeter at Mytilene: A Diachronic and Contextual Study." Institute of Archaeology, University College, 1997.

Cruz, Gaspar da. "Treatise in Which the Things of China Are Related at Great Length, with Their Particularities." In *South China in the Sixteenth Century*, edited by C.R. Boxer. Bangkok: Orchid Press, 2004.

Dalrymple, Theodore, and Trinny Woodall. "Cure Addiction the Mao Tse-Tung Way." *Spectator*, November 30, 2013.

Dalrymple, William. "The East India Company: The Original Corporate Raiders." *The Guardian*, March 4, 2015, sec. World News.

Das, N. Review of *Review of China and the Brave New World: A Study of the Origins of the Opium War 1840–42*, by Tan Chung. *Modern Asian Studies* 14, no. 3 (1980): 518–521.

Davenport-Hines, Richard. *The Pursuit of Oblivion: A Global History of Narcotics*. New York: W. W. Norton, 2002.

Davis, Angela Y. *Blues Legacies and Black Feminism*. New York: Random House–Vintage, 1999.

Davis, Kristina, and Sandra Dibble. "Fentanyl Has Taken over America's Drug Market. Where Is It Coming From?" *San Diego Union-Tribune*, June 17, 2018.

Day, Jo. "Botany Meets Archaeology: People and Plants in the Past." *Journal of Experimental Botany* 64, no. 18 (December 1, 2013): 5805–5816. doi.10.1093/jxb/ert068.

De Quincey, Thomas. *Confessions of an English Opium-Eater: Being an Extract from the Life of a Scholar*. Introduction and notes by George Armstrong Wauchope. Boston: D.C. Heath & Co., 1898. Google Books.

"Dead Man." "Trippy Films: The Wizard of Oz (1939)." *Turn Me On, Dead Man* (blog), February 17, 2013. http://turnmeondeadman.com/trippy-films-the-wizard-of-oz-1939/.

Deprez, Esmé E., Li Hui, and Ken Wills. "Deadly Chinese Fentanyl Is Creating a New Era of Drug Kingpins." *Bloomberg*, May 22, 2018, sec. Features. https://www.bloomberg.com/news/features/2018-05-22/deadly-chinese-fentanyl-is-creating-a-new-era-of-drug-kingpins.

Derks, Hans. *History of the Opium Problem: The Assault on the East, ca. 1600–1950*. Leiden, The Netherlands: Brill, 2012.

Deshpande, Jaidev. "How Was It Possible for Britain to Colonize So Many Countries?" *Quora*, January 16, 2016. https://www.quora.com/How-was-it-possible-for-Britain-to-colonise-so-many-countries.

Dickey, Colin. "The Addicted Life of Thomas De Quincey." *Lapham's Quarterly*, March 19, 2013. https://www.laphamsquarterly.org/roundtable/addicted-life-thomas-de-quincey.

Dikötter, Frank, Lars Laamann, and Zhou Xun. *Narcotic Culture: A History of Drugs in China*. London: C. Hurst, 2004.

Dolin, Eric Jay. "America's Early Trade with China." *China Business Review*, January 1, 2013.

———. "How the China Trade Helped Make America." *The Daily Beast*, November 4, 2012. https://www.thedailybeast.com/articles/2012/11/04/how-the-china-trade-helped-make-america.

———. *When America First Met China: An Exotic History of Tea, Drugs, and Money in the Age of Sail*. New York: Liveright, 2012.

Dormandy, Thomas. *Opium: Reality's Dark Dream*. New Haven, CT: Yale University Press, 2012. Kindle edition.

———. *The Worst of Evils: The Fight Against Pain*. New Haven, CT: Yale University Press, 2006.

Dorrington, K. L., and W. Poole. "The First Intravenous Anaesthetic: How Well Was It Managed and Its Potential Realized." *British Journal of Anaesthesis* 110, no. 1 (2013): 7–12. https://doi.org/10.1093/bja/aes388.

Downs, Jacques M. "American Merchants and the China Opium Trade, 1800–1840." *The Business History Review* 42, no. 4 (1968): 418–442.

"Dr. Weirde." "Chinatown's Opium Dens." *FoundSF*. Accessed June 17, 2018. http://www.foundsf.org/index.php?title=Chinatown%27s_Opium_Dens.

Drahl, Carmen. "Five Things To Know About Heroin's Curious Chemistry History." *Forbes*, January 12, 2017.

———. "What You Need To Know About Opioids, The Painkillers Behind A Crisis." *Forbes*, June 2, 2017.

Drash, Wayne, and Nadia Kounang. "Opioid Commission Member: Our Work Is a 'Sham.'" CNN, January 24, 2018. https://www.cnn.com/2018/01/23/health/patrick-kennedy-opioid-commission-sham/index.html.

"Drug Ads Gallery, 1900–1909." *ProCon*, April 7, 2014. https://prescriptiondrugs.procon.org/view.resource.php?resourceID=005589.

Drug Enforcement Administration. "DEA History—The Early Years." Accessed June 28, 2018. https://www.dea.gov/about/history.shtml.

———. "DEA Speeds up Application Process For Research On Schedule I Drugs." January 18, 2018. https://www.dea.gov/press-releases/2018/01/18/dea-speeds-application-process-research-schedule-i-drugs.

———. "Drugs of Abuse: A DEA Resource Guide." US Department of Justice, n.d. https://www.dea.gov/sites/default/files/drug_of_abuse.pdf.

"Drug Free America Foundation." DFAF. Accessed September 30, 2018. https://www.dfaf.org/.

"Drug Treatment Court Database." NDCRC.org. https://ndcrc.org/database/.

"Drug Overdose Deaths." Washington, DC: Center for Disease Control and Prevention, December 21, 2018. https://www.cdc.gov/drugoverdose/data/statedeaths.html.

Dulles, Foster Rhea. *The Old China Trade*. Edited by Bill Thayer. Cambridge, MA: Riverside Press, 1930. http://penelope.uchicago.edu/Thayer/E/Gazetteer/Places/America/United_States/_Topics/history/_Texts/DULOCT/2*.html.

"Earth." "Can Morphine Pills Be Smoked?" Ask Erowid, May 1, 2006. https://erowid.org/ask/ask.php?ID=1537.

"EBP Resource Center." Text, January 29, 2018. https://www.samhsa.gov/ebp-resource-center.

Edelman, Marian Wright. "Children and the Opioid Crisis." *Huffington Post* (blog), October 27, 2017. https://www.huffingtonpost.com/entry/children-and-the-opioid-crisis_us_59f39e1be4b06acda25f49cf.

Editors of Encyclopedia Britannica. "Aulus Cornelius Celsus—Roman Medical Writer." In *Encyclopædia Britannica*. https://www.britannica.com/biography/Aulus-Cornelius-Celsus.

Elleman, Bruce. "Historical Piracy and Its Impact." In *Histories of Transnational Crime*. Switzerland: Springer Nature, 2015.

Elliot, Captain Charles. "Cap. Charles Elliot, Chief Superintendent of Trade, to Hong Kong Residents." USC US-China Institute, February 2, 1841. https://china.usc.edu/cap-charles-elliot-chief-superintendent-trade-hong-kong-residents-february-2-1841.

Engels, Donald W. *Alexander the Great and the Logistics of the Macedonian Army*. Oakland: University of California Press, 1980.

Epimeron. "Discovery Opens Door for Synthetic Opioids with Less Addictive Qualities." *Medical Xpress*, June 1, 2018. https://medicalxpress.com/news/2018-06-discovery-door-synthetic-opioids-addictive.html.

Epstein, Edward Jay. *Agency of Fear*. Verso (Revised 1990). New York: G.P. Putnam's Sons, 1977.

Erowid Character Vaults. "Lotsof."*Erowid*. Accessed November 11, 2018. https://erowid.org/culture/characters/lotsof_howard/lotsof_howard.shtml.

Erowid, Fire, and Earth Erowid. "Erowid," n.d. https://erowid.org.

———. "Myth of the 'Hashish' / 'Assassin' Connection." *Erowid Cannabis Vault* (blog). Accessed February 5, 2018. https://erowid.org/plants/cannabis/cannabis_info4.shtml.

Escohotado, Antonio. *A Brief History of Drugs: From the Stone Age to the Stoned Age.* Rochester, VT: Park Street Press, 1999. Kindle edition.

———. *The General History of Drugs.* Vol. 1. Valparaiso, Chile: Graffiti Militante Press, 2010. Google Books.

European Food Safety Authority. "Opium Alkaloids in Poppy Seeds: Assessment Updated." *EFSA*, May 18, 2018. https://www.efsa.europa.eu/en/press/news/180516.

Eustachewich, Lia. "El Chapo's Mexico Drug Cartel Is Doing Just Fine without Him." *New York Post* (blog), November 13, 2018.

Everett, Nicholas. *The Alphabet of Galen: Pharmacy from Antiquity to the Middle Ages.* Toronto: University of Toronto Press, 2012.

Fahey, David M., and Jon S. Miller, eds. *Alcohol and Drugs in North America: A Historical Encyclopedia.* Vol. 1. Santa Barbara, CA: Gale Virtual Reference Library, 2013.

"FastStats." Centers for Disease Control and Prevention, March 17, 2017. https://www.cdc.gov/nchs/fastats/marriage-divorce.htm.

Felbab-Brown, Vanda. "Afghanistan's Opium Production Is through the Roof—Why Washington Shouldn't Overreact." *Brookings* (blog), November 21, 2017. https://www.brookings.edu/blog/order-from-chaos/2017/11/21/afghanistans-opium-production-is-through-the-roof-why-washington-shouldnt-overreact/.

———. "The Poppy Problem." *Brookings* (blog), September 21, 2007. https://www.brookings.edu/opinions/the-poppy-problem/.

Fell-Smith, Charlotte, and Royal College of Physicians of London. *John Dee, 1527–1608.* London: Constable & Company, 1909. Internet Archive.

Ferreira, Susana. "Portugal's Radical Drugs Policy Is Working. Why Hasn't the World Copied It?" *The Guardian*, December 5, 2017, sec. News.

Ferreiro, Carmen. *Heroin.* Drugs—The Straight Facts. New York: Chelsea House, 2003.

Feuer, Alan. "El Chapo Found Guilty on All Counts; Faces Life in Prison." *New York Times*, February 12, 2019. http://www.nytimes.com/2019/02/12/nyregion/el-chapo-verdict.html.

Filan, Kenaz. *The Power of the Poppy: Harnessing Nature's Most Dangerous Plant Ally.* Rochester, VT: Park Street Press, 2011. Kindle edition.

"Five Genius Minds Who Dabbled in Drugs." *DrugAbuse* (blog), November 20, 2014. https://drugabuse.com/5-genius-minds-who-dabbled-in-drugs/.

Flascha, Carlo. "On Opium: Its History, Legacy and Cultural Benefits." Prospect Journal of International Affairs at UCSD, May 25, 2011. https://prospectjournal.org/2011/05/25/on-opium-its-history-legacy-and-cultural-benefits/.

Fleming, Nic. "The Search for the Perfect Painkiller." *The Observer,* January 28, 2018, sec. US news.

Fogoros, Richard N., and "Buddy T." "The Costs of Drug Use to Society." *Verywell Mind,* June 16, 2018. https://www.verywellmind.com/what-are-the-costs-of-drug-abuse-to-society-63037.

Forbes, Daniel. "Prime-Time Propaganda." *Salon,* January 13, 2000. https://www.salon.com/2000/01/13/drugs_6/.

Frakt, Austin. "Opioid Use and Policy in the US—a History." *The Incidental Economist* (blog), December 24, 2014. https://theincidentaleconomist.com/wordpress/opioid-use-and-policy-in-the-us-a-history/.

France, Duncan. "China Receives Letter from Queen Elizabeth 1—383 Years Too Late." *New Zealand China Friendship Society* (blog), August 6, 2016. http://nzchinasociety.org.nz/?p=26466.

Freud, Sigmund. *Moses and Monotheism.* London: Hogarth Press, 1939.

Friedman, Richard. "Feel-Good Gene." *New York Times.* March 6, 2015, sec. Opinion.

Gabriel, Richard A. *The Madness of Alexander the Great: And the Myth of Military Genius.* South Yorkshire, England: Pen & Sword Military, 2015. Kindle edition.

"Gene Therapy for Addiction." *DrugRehab* (blog), May 16, 2017. https://www.drugrehab.org/gene-therapy-for-addiction/.

Gerth, Stephen Engelberg with Jeff Gerth, and Special to the New York Times. "Bush and Noriega: Examination of Their Ties." *New York Times,* September 28, 1988, sec. U.S. https://www.nytimes.com/1988/09/28/us/bush-and-noriega-examination-of-their-ties.html.

Ghafour, Hamida. "Afghan Farmers Turn to Saffron as Replacement for Their Opium Crops." *The Telegraph,* April 4, 2004.

Gieringer, Dale. "125 Years of the War on Drugs." DrugSense—Media Awareness Project, November 2000. http://www.drugsense.org/dpfca/opiumlaw.html.

Golshani, Seyyed Alirez. "Hakim Imad Al-Din Mahmud Ibn-Mas'ud Shirazi (1515–1592), a Physician and Social Pathologist of Safavid Era." *Galen Medical Journal* 2, no. 4 (2013): 169–173.

Goodman, Susan. "Israel's Mr. Pomegranate." *Israel21c*, September 18, 2015. http://www.israel21c.org/israels-mr-pomegranate/.

Grant, Frederic Delano Jr. "Six Years of Rebound and Opium, 1815–1828." In *The Chinese Cornerstone of Modern Banking*. Studies in the History of Private Law 6. Edited by Frederic Delano Jr. Brill Online Books and Journals, 2014. doi.10.1163/9789004276567_007.

Gray, Richard. "Ancient Humans Were Taking Drugs up to 10,600 Years Ago." *Daily Mail*, February 10, 2015.

Green, Jacob. "An Address on the Botany of the United States: Delivered Before the Society for the Promotion of Useful Arts. To Which Is Added A Catalogue of Plants Indigenous to the State of New York." Websters and Skinners, 1814. Google Books.

Greenberg, Andy. "Silk Road Creator Ross Ulbricht Loses His Life Sentence Appeal." *Wired*, May 31, 2017. https://www.wired.com/2017/05/silk-road-creator-ross-ulbricht-loses-life-sentence-appeal/.

Griffenhagen, George B., and James Harvey Young. "Old English Patent Medicines in America." Project Gutenberg EBook, 2009. http://www.gutenberg.org/cache/epub/30162/pg30162.txt.

Griner, Allison. "China's New Opium Wars: Battling Addiction in Beijing." *Al Jazeera*, July 6, 2016. https://www.aljazeera.com/indepth/features/2016/05/china-opium-wars-battling-addiction-beijing-160516141819379.html.

Grout, James. "Nicander." In *Encyclopaedia Romana*. University of Chicago, n.d. Accessed March 27, 2018.

Guerra-Doce, Elisa. "Psychoactive Substances in Prehistoric Times: Examining the Archaeological Evidence." *Time and Mind* 8, no. 1 (January 2, 2015): 91–112. doi.10.1080/1751696X.2014.993244.

Gunn, Geoffrey. *World Trade Systems of the East and West*. Leiden, Netherlands: Brill, 2017.

Haggard, Howard W. *Devils, Drugs, and Doctors: The Story of the Science of Healing from Medicine-Man to Doctor*. London: William Heinemann, 1913.

Hajar, Rachel. "The Air of History: Early Medicine to Galen (Part I)." *Heart Views: The Official Journal of the Gulf Heart Association* 13, no. 3 (2012): 120–128. doi.10.4103/1995-705X.102164.

Hall, Captain W. H., and W. D. Bernard. *The Nemesis in China, Comprising a History of the Late War in That Country*. 3rd ed. London: Henry Colburn, 1846.

Halpern, John H. "Addiction Is a Disease." *Psychiatric Times* 19, no. 10 (October 1, 2002).

———. "The Right Medicine for World Drug Day." *Indian Journal of Medical Research* 135, no. 6 (June 2012): 801–802.

———. "The Use of Hallucinogens in the Treatment of Addiction." *Addiction Research* 4, no. 2 (1996): 177–189.

Hamilton, Megan. "Five Fascinating Facts You Might Not Know about Sea Otters." *Digital Journal,* October 12, 2015. http://www.digitaljournal.com/news/environment/five-fascinating-facts-you-might-not-know-about-sea-otters/article/446219.

Hancock, Shawn. "Hancock Shaker Village FAQs." Hancock Shaker Village, 2018. https://hancockshakervillage.org/shakers/shaker-history-faqs/.

Hansson, Maria C., and Brendan P. Foley. "Ancient DNA Fragments inside Classical Greek Amphoras Reveal Cargo of 2400-Year-Old Shipwreck." *Journal of Archaeological Science* 35, no. 5 (May 1, 2008): 1169–1176. doi.10.1016/j.jas.2007.08.009.

Haqqi, Ty. "Eight Countries That Produce the Most Weed in the World." *Insider Monkey,* May 9, 2016. http://www.insidermonkey.com/blog/8-countries-that-produce-the-most-weed-in-the-world-447869/.

Hari, Johann. *Chasing the Scream: The First and Last Days of the War on Drugs.* New York: Bloomsbury, 2015.

———. "The Hunting of Billie Holiday." *Politico Magazine,* January 17, 2015. https://www.politico.com/magazine/story/2015/01/drug-war-the-hunting-of-billie-holiday-114298#.VOoZ2i5X_Yh.

Harper, Douglas. "Junk." *Online Etymology Dictionary* (blog), 2018. https://www.etymonline.com/word/junk.

Harrison Act, Pub. L. No. H.R. 6282 (1914).

"Harrison Drug Act Takes Effect March 1." *Richmond Times-Dispatch,* February 17, 1915.

Harrison, Lana D., Michael Backenheimer, and James A. Inciardi. "History of Drug Legislation." In *Cannabis Use in the United States,* 237–247. Amsterdam: University of Amsterdam, 1999.

Harwood, Matthew. "One Thing You Can Say for the War on Drugs…Is We Gave It a Fair Shot.'" American Civil Liberties Union, April 23, 2015. https://www.aclu.org/blog/criminal-law-reform/drug-law-reform/one-thing-you-can-say-war-drugs-we-gave-it-fair-shot.

Hayes, Jack Patrick. "The Opium Wars in China." Asia Pacific Foundation of Canada, 2017. https://asiapacificcurriculum.ca/system/files/2017-11/The%20Opium%20Wars%20in%20China%20-%20Background%20Reading_1.pdf.

Hays, Jeffrey. "Ancient Roman Food, Spices, and Banquets." Facts and Details, 2008. http://factsanddetails.com/world/cat56/sub369/item2071.html.

———. "Assassins, Hashish and Ninjas." Facts and Details, July 2008. http://factsanddetails.com/world/cat58/sub385/item2375.html.

Heard, Albert. "A Chronicle of the China Trade. The Papers of Augustine Heard & Co., 1840–1877." Harvard Business School: Baker Library Historical Collections Exhibit, January 23, 1856. https://www.library.hbs.edu/hc/heard/index.html.

Hedman, Jason. "Commissioner Lin Zexu and the Opium War." *Jasonhedman* (blog), November 28, 2012. https://jasonhedman.wordpress.com/2012/11/28/commissioner-lin-zexu-and-the-opium-war/.

Henn, Debra, and Deborah DeEugenio. *Barbiturates*. Drugs—The Straight Facts. New York: Chelsea House, 2007.

Hern, Alex. "Five Stupid Things Dread Pirate Roberts Did to Get Arrested." *The Guardian*, October 3, 2013, sec. Technology.

"Heroin-Assisted Treatment in Switzerland: Successfully Regulating the Supply and Use of a High-Risk Injectable Drug." *Transform: Getting Drugs Under Control* (blog), January 10, 2017. https://www.tdpf.org.uk/blog/heroin-assisted-treatment-switzerland-successfully-regulating-supply-and-use-high-risk-0.

Heydari, Mojtaba, M. Hashempur, and Arman Zargaran. "Medicinal Aspects of Opium as Described in Avicenna's Canon of Medicine." *Acta Medico-Historica Adriatica: AMHA* 11 (July 25, 2013): 101–112.

Higgins, Tucker. "Trump Calls on China to Seek Death Penalty for Fentanyl Distributors." CNBC, December 5, 2018. https://www.cnbc.com/2018/12/05/trump-calls-on-china-to-seek-death-penalty-for-fentanyl-distributors.html.

Hilgers, Laura. "Treat Addiction Like Cancer." *New York Times*, May 19, 2018, sec. Opinion.

Hillman, David. *The Chemical Muse: Drug Use and the Roots of Western Civilization*. New York: Thomas Dunne, 2008.

Hippocrates. *Hippocrates Volume 1*. Translated by W. H. S. Jones. Harvard Classics. Cambridge, MA: Harvard University Press, 1923.

Hirst, K. Kris. "Can We Blame Europeans for the Domestication of Opium Poppy?" *ThoughtCo*, March 8, 2017. https://www.thoughtco.com/opium-poppy-the-history-of-domestication-169375.

———. "Who Were the First Farmers of Europe and How Did They Get There?" *ThoughtCo*, March 8, 2017. https://www.thoughtco.com/linearbandkeramik-culture-farming-innovators-171552.

History of Humanity. "Indian Ocean." Golden Age of Piracy. Accessed February 26, 2018. http://www.goldenageofpiracy.org/golden-age-of-piracy/indian-ocean.php.

"The History of Injecting, and the Development of the Syringe." Exchange

Supplies. Accessed May 26, 2018. http://www.exchangesupplies.org/ article_history_of_injecting_and_development_of_the_syringe.php.

"History of Macau as a Portuguese Colony." *Yesterday's Shadow* (blog), December 2, 2012. https://yesterdaysshadow.wordpress.com/history-of-macau-as-a-portuguese-colony/.

"A History of Opiate Laws in the United States." The National Alliance of Advocates for Buprenorphine Treatment. Accessed June 13, 2018. https://www.naabt.org/laws.cfm.

"The History of the Drug Hydrocodone." *A Forever Recovery* (blog), 2018. http://aforeverrecovery.com/drug-addiction/painkiller/the-history-of-hydrocodone/.

Hodgson, Barbara. *Opium: A Portrait of the Heavenly Demon.* San Francisco: Chronicle Books, 1999.

Hogshire, Jim. *Opium for the Masses: Harvesting Nature's Best Pain Medicine.* Port Townsend, WA: Feral House, 2009. Kindle edition.

Holman, John. "El Chapo Trial: It's Business as Usual for Sinaloa Drug Cartel." *Mexico News—Al Jazeera,* November 4, 2018. https://www.aljazeera.com/news/2018/11/el-chapo-trial-its-business-usual-sinaloa-drug-cartel-181104105058455.html.

Huard, Pierre, and Ming Wong. *Chinese Medicine.* World University Library. New York: McGraw-Hill, 1968.

Hunter, Frances. "Lewis & Clark's Medicine Chest." *Frances Hunter's American Heroes* (blog), February 18, 2010. https://franceshunter.wordpress.com/2010/02/18/lewis-clarks-medicine-chest/.

"India Opium." *American Druggist* 20, no. 22 (November 16, 1891). https://books.google.com/books?id=Z1BHAQAAMAAJ&pg=PA344&dq =American+Druggist+India+Opium&hl=en&sa=X&ved=0ahUKEwiwv O_HjbneAhXJoFMKHXGBBOgQ6AEIKjAA#v=onepage&q=American %20Druggist%20India%20Opium&f=false.

Inglis, Brian. *The Forbidden Game: A Social History of Drugs.* Great Britain: Hodder & Stoughton, 1975.

"Inmate 'Accidentally' Died after Smoking Morphine Patch." *BBC News,* January 31, 2018. http://www.bbc.com/news/uk-england-essex-42894468.

Institute for Research, Education & Training in Addictions. "A Look at Treatment History: The Narcotic Farm." IRETA. Accessed November 12, 2018. https://ireta.org/resources/a-look-at-treatment-history-the-narcotic-farm/.

Invaluable (Auction House). "Morris, Robert. Autograph Letter Signed."

Autographed Letter, May 1776. https://www.invaluable.com/auction-lot/morris-robert-autograph-letter-signed-1-may-17-137-c-1f14b0a946.

Isacson, Adam. "Four Common Misconceptions about U.S.-Bound Drug Flows through Mexico and Central America." WOLA, June 20, 2017. https://www.wola.org/analysis/four-common-misconceptions-u-s-bound-drug-flows-mexico-central-america/.

Islamic Republic of Afghanistan Ministry of Counter Narcotics. "Afghanistan Opium Survey 2017: Cultivation and Production." New York: United Nations Office on Drugs & Crime, November 2017.

Itkowitz, Colby. "Senate Easily Passes Sweeping Opioids Legislation, Sending It to President Trump." *Washington Post*, October 13, 2018.

Jacobson, Katherine. "Silk Road 101: How Did the Now-Busted Online Black Market Work?" *Christian Science Monitor,* October 3, 2013.

Jaffe, Bernard. *Crucibles: The Story of Chemistry from Ancient Alchemy to Nuclear Fission.* Dover, 1976. First published 1930 by Simon & Schuster.

Jay, Mike. *Emperors of Dreams: Drugs in the Nineteenth Century.* 2nd ed. Cambs, England: Daedelus, 2011. Kindle edition.

Jenkins, P. Nash. "Heroin Addiction's Fraught History." *The Atlantic,* February 24, 2014.

Jensen, Joan M. *With These Hands: Women Working on the Land.* Old Westbury, NY: Feminist Press, 1981.

Johns, Timothy. *The Origins of Human Diet and Medicine: Chemical Ecology.* Tucson: University of Arizona Press, 1996

Jones, Adam James. "Bad Medicine: A History of Narcotics in Pharmaceuticals." *Rocky Mountain Legends* (blog), July 11, 2012. https://adamjamesjones.wordpress.com/2012/07/11/bad-medicine-a-history-of-narcotics-in-medicine/.

"Judge Hull's Death from an Overdose of Morphine." *Omaha Daily Bee,* March 4, 1887, sec. Local News Budget.

Kane, Harry Hubbell. *Drugs That Enslave: The Opium, Morphine, Chloral and Hashisch Habits.* Philadelphia: P. Blakiston, 1881. HathiTrust. https://catalog.hathitrust.org/Record/001581485.

———. *Opium-Smoking in America and China: A Study of Its Prevalence, and Effects Immediate and Remote, on the Individual and the Nation.* New York: G.P. Putnam's Sons, 1882.

Kaplan, Ezra. "A Growing Number of Sober Programs Support College Students Recovering from Addiction." *NBC News.* Accessed March 19, 2019. https://www.nbcnews.com/news/us-news/growing-number-sober-programs-support-college-students-recovering-addiction-n875326.

Keay, John. *The Honourable Company: A History of the English East India Company.* London: Macmillan, 1994. Google Books.

Keeling, Julian. "The Drugs Don't Work. Review: 'Opium: A Portrait of the Heavenly Demon,' Barbara Hodgson." *New Statesman,* December 4, 2000.

Kennedy, Patrick J. "Recommendations of Congressman Patrick J. Kennedy to the President's Commission on Combating Drug Addiction and the Opioid Crisis." Kennedy Forum, October 2017. https://www.thekennedyforum.org/congressman-patrick-j-kennedy-issues-recommendations-to-the-presidents-commission-on-combating-drug-addiction-and-the-opioid-crisis/.

———. "Revolutionizing and Standardizing Mental Healthcare." Kennedy Forum. Accessed September 30, 2018. https://www.thekennedyforum.org/.

Khazan, Olga. "The Surprising Ease of Buying Deadly Drugs Online." *The Atlantic,* January 3, 2018.

Kienholz, Mary. *Opium Traders and Their Worlds. A Revisionist Exposé of the World's Greatest Opium Traders.* Vol. 1. New York: iUniverse, 2008.

———. *Opium Traders and Their Worlds: A Revisionist Exposé of the World's Greatest Opium Traders.* Vol. 2. New York: iUniverse, 2008.

King, Hobart M. "Uses of Gold in Industry, Medicine, Computers, Electronics, Jewelry." Geology, 2018. https://geology.com/minerals/gold/uses-of-gold.shtml.

King, Rufus. "Dr. Ratigan's Lonely Battle (Ch. 7)." In *The Drug Hang Up: America's Fifty-Year Folly.* Springfield, IL: Charles C. Thomas, 1974.

Kissinger, C. Clark. "How Maoist Revolution Wiped out Drug Addiction in China." Revcom. Accessed April 14, 2018. http://www.revcom.us/a/china/opium.htm.

Koon, Wee Kek. "How Tobacco First Reached China, 450 Years Before the Country Became the World's Largest Consumer." *South China Morning Post,* December 15, 2016. http://www.scmp.com/magazines/post-magazine/short-reads/article/2054796/how-tobacco-first-reached-china-450-years.

Korsemeyer, Pamela, and Henry R. Kranzer, eds. *Encyclopedia of Drugs, Alcohol & Addictive Behavior.* 3rd ed. Vol. 3. Macmillan Reference USA, 2009.

Kounang, Nadia. "Fentanyl Is the Deadliest Drug in America, CDC Confirms." CNN. Accessed December 12, 2018. https://www.cnn.com/2018/12/12/health/drugs-overdose-fentanyl-study/index.html.

Kramer, John C. "Speculations on the Nature and Pattern of Opium

Smoking." *Journal of Drug Issues* 9, no. 2 (April 1, 1979): 247–255. doi.10.1177/002204267900900209.

Krishnamurti, Chandrasekhar, and SSC Chakra Rao. "The Isolation of Morphine by Sertürner." *Indian Journal of Anaesthesia* 60, no. 11 (November 2016): 861–862. doi.10.4103/0019-5049.193696.

Kristof, Nicholas. "How to Win a War on Drugs." *New York Times*, September 22, 2017, sec. Opinion.

Kritikos, P. G., and S. P. Papadaki. "The History of the Poppy and of Opium and Their Expansion in Antiquity in the Eastern Mediterranean Area." *Journal of the Archaeological Society of Athens*, January 1, 1967: 17–38.

Kunzig, Robert, and Jennifer Tzar. "La Marmotta." *Discover Magazine*, November 1, 2002.

Kurtzman, Daniel. "What Are the Most Ridiculous Richard Nixon Quotes?" ThoughtCo, December 29, 2018. https://www.thoughtco.com/richard-nixon-quotes-2733879.

Lalanne, Laurence, Chloe Nicot, Jean-Philippe Lang, Gilles Bertschy, and Eric Salvat. "Experience of the Use of Ketamine to Manage Opioid Withdrawal in an Addicted Woman: A Case Report." *BMC Psychiatry* 16 (October 11, 2016). doi.10.1186/s12888-016-1112-2.

Latimer, Dean, and Jeff Goldberg. *Flowers in the Blood: The Story of Opium.* New York: Franklin Watts, 1981.

Lau, Jolene L., and Michael K. Dunn. "Development Trends for Peptide Therapeutics: Status in 2016." Presented at the Annual Peptide Therapeutics Symposium, La Jolla, CA, October 27, 2016. http://www.peptidetherapeutics.org/wp-content/uploads/2017/02/2016-Peptide-Therapeutics-Poster-Ferring-Research-Institute.pdf.

Lavin, Arthur. "The American Opium Epidemic." *Advanced Pediatrics*, April 18, 2017.

Lazich, Michael C. "American Missionaries and the Opium Trade in Nineteenth-Century China." *Journal of World History* 17, no. 2 (June 2006): 197–223.

Lazzara, Liz. "Ten Ways to Get off Opiates." Recovery Village. Accessed September 14, 2018. https://www.therecoveryvillage.com/opiate-addiction/how-to-get-off-opiates/.

Lee, Seungyeop, and Dong-Kwon Rhee. "Effects of Ginseng on Stress-Related Depression, Anxiety, and the Hypothalamic-Pituitary-Adrenal Axis." *Science Direct: Journal of Ginseng Research* 41, no. 4 (October 2017): 589–594.

Levenick, Christopher. "The Philanthropy Hall of Fame." Philanthropy

Roundtable. Accessed May 11, 2018. http://www.philanthropyroundtable.
org/almanac/hall_of_fame/thomas_perkins.

Lewin, Louis. *Phatastica: A Classic Survey on the Use and Abuse of Mind-Altering Plants*. Translated from second German edition by P. H. A. Wirth. Rochester, VT: Inner Traditions, 1998.

Lewis, Jone Johnson. "A Surprisingly Long List of Medieval Women Writers." ThoughtCo, June 29, 2107. https://www.thoughtco.com/medieval-women-writers-3530911.

Libby, Ronald T. "The DEA's War on Doctors: A Surrogate for the War on Drugs." Congressional Briefing on The Politics of Pain: Drug Policy & Patient Access to Effective Pain Treatments (2004). http://www.aapsonline.org/painman/paindocs2/libbystatement.pdf.

———. "Treating Doctors as Drug Dealers." *Policy Analysis*, no. 545 (June 16, 2005): 28.

"Linder v. United States—US Supreme Court." Schaffer Library of Drug Policy, April 13, 1925. http://druglibrary.net/schaffer/History/linderv.htm.

Lingle, Brett, Carolyn Kousky, and Leonard Shabman. "Federal Disaster Rebuilding Spending: A Look at the Numbers." *Risk Management and Decision Processes Center* (blog), February 22, 2018. https://riskcenter.wharton.upenn.edu/disaster-aid/federal-disaster-rebuilding-spending-look-numbers/.

Lonely Planet. "History of Hong Kong," 2018. https://www.lonelyplanet.com/china/hong-kong/history#pageTitle.

Lopez, German. "How Much Does the War on Drugs Cost?" *Vox*, August 21, 2014. https://www.vox.com/cards/war-on-drugs-marijuana-cocaine-heroin-meth/war-on-drugs-cost-spending.

———. "How Obama Quietly Reshaped America's War on Drugs." *Vox*, December 19, 2016. https://www.vox.com/identities/2016/12/19/13903532/obama-war-on-drugs-legacy.

———. "Trump Just Signed a Bipartisan Bill to Confront the Opioid Epidemic." *Vox*, September 28, 2018. https://www.vox.com/policy-and-politics/2018/9/28/17913938/trump-opioid-epidemic-congress-support-act-bill-law.

———. "We Looked for a State That's Taken the Opioid Epidemic Seriously. We Found Vermont." *Vox*, October 30, 2017. https://www.vox.com/policy-and-politics/2017/10/30/16339672/opioid-epidemic-vermont-hub-spoke.

Lovell, Julia. *The Opium War: Drugs, Dreams and the Making of China*. London: Picador, 2011.

Lyons, Albert. "Ancient China: Health Guidance." Health Guidance, 2018. http://www.healthguidance.org/entry/6333/1/Ancient-China.html.

Macht, David I. "The History of Opium and Some of Its Preparations and Alkaloids." *Journal of the American Medical Association* 64, no. 6 (February 6, 1915): 477–481. doi.10.1001/jama.1915.02570320001001.

MacLeod, Calum. "Executions Hit New High in China's Drugs War." *The Independent*, June 28, 2000.

Macy, Beth. "'I Am Going to Die If I Keep Living the Way I Am.' She Was Right." *New York Times*, July 20, 2018, sec. Opinion.

Makepeace, Margaret. "A Brief History of the English East India Company 1600–1858." Qatar Digital Library, August 13, 2014. https://www.qdl.qa/en/brief-history-english-east-india-company-1600%E2%80%931858.

"Manchu Emperor Daoguang / Taoukwang / Tao Kwang—1820–1850." Global Security. Accessed April 20, 2018. https://www.globalsecurity.org/military/world/china/emperor-manchu-daoguang.htm.

Mann, M. J. "Alexander's Injuries Part 1." *Second Achilles* (blog), October 25, 2013. https://thesecondachilles.com/2013/10/25/alexanders-injuries-part-1/.

"Marco Polo Introduces Opium to the West." This Day in Alternate History, May 4, 2015. http://thisdayinalternatehistory.blogspot.com/2015/05/guest-post-marco-polo-introduces-opium.html.

Margotta, Roberto. *The Story of Medicine*. New York: Golden Press, 1967.

Maron, Dina Fine. "How Opioids Kill." *Scientific American*, January 8, 2018.

Marshall, Jonathan. "Cooking the Books: The Federal Bureau of Narcotics, the China Lobby and Cold War Propaganda, 1950–1962." *Asia-Pacific Journal: Japan Focus* 11, no. 37 (September 15, 2013).

Martin, Steven. "How Collecting Opium Antiques Turned Me Into an Opium Addict." Interview by Lisa Hix, September 24, 2012. https://www.collectorsweekly.com/articles/journey-into-the-opium-underworld/.

———. *Opium Fiend: A 21st Century Slave to a 19th Century Addiction*. New York: Villard—Random House, 2012.

———. "Opium Museum." Opium Museum, 2007. http://www.opiummuseum.com/index.pl?introduction.

"Mass General History." Massachusetts General Hospital, 2018. https://www.massgeneral.org/museum/history/default.aspx.

McCoy, Alfred W. "How the Heroin Trade Explains the US-UK Failure in Afghanistan." *The Guardian*, January 8, 2018.

McGreevy, Patrick. "As the Top Pot-Producing State in the Nation, California

Could Be on Thin Ice with the Federal Government." *LA Times*, October 1, 2017.

McKay, Brett, and Kate McKay. "Male Rites of Passages from Around the World." *Art of Manliness* (blog), February 21, 2010. https://www.artofmanliness.com/articles/male-rites-of-passage-from-around-the-world/.

McKenna, Terence. *Food of the Gods: The Search for the Original Tree of Knowledge*. New York: Bantam Books, 1992.

McLean Hospital. "Expert Treatment for Alcohol and Drug Abuse at Harvard Affiliated Hospital." Accessed May 14, 2018. https://www.mcleanhospital.org/programs/alcohol-and-drug-abuse-inpatient-program.

"McLean Hospital." In *American Hospital Directory*, March 2, 2018. https://www.ahd.com/free_profile/224007/McLean_Hospital/Belmont/Massachusetts/.

McWilliams, John C. *The Protectors: Harry J. Anslinger and the Federal Bureau of Narcotics, 1930–1962*. Cranbury, NJ: Associated University Presses, 1990. Internet Archive.

———. "Unsung Partner Against Crime: Harry J. Anslinger and the Federal Bureau of Narcotics, 1930–1962." *Pennsylvania Magazine of History and Biography* 113, no. 2 (1989): 207–236.

Meier, Barry. "Origins of an Epidemic: Purdue Pharma Knew Its Opioids Were Widely Abused." *New York Times*, May 29, 2018.

———. "In Guilty Plea, OxyContin Maker to Pay $600 Million." The *New York Times*, May 10, 2007. https://www.nytimes.com/2007/05/10/business/11drug-web.html.

Merlin, Mark David. *On the Trail of the Ancient Opium Poppy*. London: Associated University Presses, 1984.

Merrilles, R. S. "Opium Trade in the Bronze Age Levant." *Antiquity* 36 (1962): 287–292.

Miller, Amy Bess. *Shaker Herbs: A History and a Compendium*. New York: Clarkson N. Potter, 1976.

Miller, M. Stephen. *From Shaker Lands and Shaker Hands*. Lebanon, NH: University Press of New England, 2007.

Mirsky, Jonathan. "The Truth about Mao. Mass Murderer, Womaniser, Liar, Drug Baron: Book Paints Horrific Portrait." Review of *Mao: The Unknown Story* by Jung Chang and Jon Halliday. *The Independent*, May 29, 2005.

MIT OpenCourseWare. "The Rise & Fall of the Canton Trade System." Visualizing Culture, 2009. https://ocw.mit.edu/ans7870/21f/21f.027/rise_fall_canton_01/pdf/cw_essay.pdf.

Monahan, John. *They Called Me Mad: Genius, Madness and the Scientists Who Pushed the Outer Limits of Knowledge.* New York: Penguin, 2010.

Montag, Guy. "How to Make Opium from a Papaver Rhoeas, or Common Poppy/Corn Poppy." Reddit, 2016. https://www.reddit.com/r/Drugs/comments/4py8i3/how_to_make_opium_from_a_papaver_rhoeas_or_common/.

Monteleone, Davide. "A New Silk Road." *New Yorker,* January 8, 2018.

Morrell, Alex. "The OxyContin Clan: The $14 Billion Newcomer to Forbes 2015 List of Richest U.S. Families." Forbes, July 1, 2017. https://www.forbes.com/sites/alexmorrell/2015/07/01/the-oxycontin-clan-the-14-billion-newcomer-to-forbes-2015-list-of-richest-u-s-families/.

Mornbelli, Armando. "Lake Dwellings Reveal Hidden Past." Swiss Info, November 11, 2016. https://www.swissinfo.ch/eng/lake-dwellings-reveal-hidden-past/30542748.

Morone, Natalia E., and Debra K. Weiner. "Pain as the 5th Vital Sign: Exposing the Vital Need for Pain Education." *Clinical Therapeutics* 35, no. 11 (November 2013): 1728–1732. doi.10.1016/j.clinthera.2013.10.001.

Moskowitze, Peter, and the Fusion Investigative Team. "Last Chances in the Second City." Fusion, 2016. http://interactive.fusion.net/death-by-fentanyl/last-chances-in-the-second-city.html.

Mukherjee, Siddhartha. *The Emperor of All Maladies: A Biography of Cancer.* New York: Scribner/Simon & Schuster, 2010.

Murray, John P. and David W. Walters. "Kill All the Poppies." *Foreign Policy* (blog), November 22, 2017. https://foreignpolicy.com/2017/11/22/kill-all-the-poppies-afghanistan-heroin-taliban/.

Musto, David F. *The American Disease: Origins of Narcotic Control.* 3rd ed. New York: Oxford University Press, 1999.

Nabiyeva, Komila. "Uzbekistan Rediscovers Lost Culture in the Craft of Silk Road Paper Makers." *The Guardian,* June 2, 2014.

Nahid, Babak. "Review of The Assassin Legends: Myths of the Isma'ilis by Farhad Daltary." Heritage Society, March 4, 2003. http://ismaili.net/Source/fd0328d.html.

National Academies of Sciences, Engineering, Health and Medicine Division, Board on Health Sciences Policy, Committee on Pain Management and Regulatory Strategies to Address Prescription Opioid Abuse. Jonathan K. Phillips, Morgan A. Ford, and Richard J. Bonnie. *Trends in Opioid Use, Harms, and Treatment: Balancing Societal and Individual Benefits and Risks of Prescription Opioid Use.* Washington, DC: National Academies Press (US), 2017.

National Center for Biotechnology Information. "Fentanyl | C22H28N2O—CID=3345." PubChem: Open Chemistry Database, n.d. Accessed August 25, 2018.

National Conference of State Legislatures. "Drug Overdose Immunity and Good Samaritan Laws." Accessed January 17, 2019. http://www.ncsl.org/research/civil-and-criminal-justice/drug-overdose-immunity-good-samaritan-laws.aspx.

National Institute on Drug Abuse. "Dramatic Increases in Maternal Opioid Use and Neonatal Abstinence Syndrome." January 22, 2019. https://www.drugabuse.gov/related-topics/trends-statistics/infographics/dramatic-increases-in-maternal-opioid-use-neonatal-abstinence-syndrome.

———. "Opiates Binding to Opiate Receptors in the Nucleus Accumbens: Increased Dopamine Release." In *The Neurobiology of Drug Addiction*. Introduction to the Brain, Section 3; Part 2. NIH, 2007. https://www.drugabuse.gov/publications/teaching-packets/neurobiology-drug-addiction/section-iii-action-heroin-morphine/4-opiates-binding-to-opiate-rece.

———. "Overdose Death Rates." DrugAbuse, August 9, 2018. https://www.drugabuse.gov/related-topics/trends-statistics/overdose-death-rates.

———. *Principles of Drug Addiction Treatment: A Research-Based Guide*. 3rd ed. 12–4189. National Institutes of Health, 2012.

National Media Affairs Office. "DEA Speeds Up Application Process for Research on Schedule I Drugs." United States Drug Enforcement Administration, January 18, 2018. https://www.dea.gov/press-releases/2018/01/18/dea-speeds-application-process-research-schedule-i-drugs.

"Neonatal Abstinence Syndrome and Opioid Policy." Accessed January 18, 2019. https://www.vumc.org/nas/.

Nevius, James. "The Strange History of Opiates in America: From Morphine for Kids to Heroin for Soldiers." *The Guardian*, March 15, 2016, sec. US news.

"New 'Dope' Law to Lessen Crime: Anti-Narcotic Act Limits Sales of All Harmful Drugs." *Lafayette Tippecanoe County Democrat*, February 26, 1915, Newspaper Archives edition.

Nicander of Colophon. *Poems and Poetical Fragments*. Edited by A. S. F. Gow and A. F. Scholfield. Cambridge: Cambridge University Press, 1953. Google Books.

Nielsen, Nick. "What Did Aristotle Teach Alexander the Great?" Quora,

2017. https://www.quora.com/What-did-Aristotle-teach-Alexander-the-Great.

Noriega, Ambassador Roger F. "Columbia: Peace with Security: Targeting Coca and Transnational Crime." Senate Caucus on International Narcotics Control (2017). https://www.drugcaucus.senate.gov/sites/default/files/Noriega%20Colombia%20Senate%20Drug%20Caucus%20091 217%20FINAL.pdf.

Norn, Svend, Poul R. Kruse, and Edith Kruse. "On the History of Injection." *Dansk Medicinhistorisk Arbog* 34 (2006): 104–113.

Norton, Stata. "The Pharmacology of Mithridatum: A 2000-Year-Old Remedy." *Molecular Interventions* 6, no. 2 (April 1, 2006): 60. doi.10.1124/mi.6.2.1.

NPR Staff. "Legalize All Drugs? The 'Risks Are Tremendous' Without Defining the Problem." *Weekend Edition.* NPR, March 27, 2016. https://www.npr.org/2016/03/27/472023148/legalize-all-drugs-the-risks-are-tremendous-without-defining-the-problem.

Nunn, Nathan, and Nancy Quian. "The Columbian Exchange: A History of Disease, Food, and Ideas." *Journal of Economic Perspectives* 24, no. 2 (Spring 2010): 163–188.

O'Brien, Patrick, ed. *Atlas of World History.* Concise Edition. Oxford: Oxford University Press, 2002.

O'Connor, J. J., and E. F. Robertson. "Avicenna Biography." JOC/EFR, November 1999. http://www-history.mcs.st-andrews.ac.uk/Biographies/Avicenna.html.

Office of Justice Programs. "Do Drug Courts Work? Findings from Drug Court Research." National Institute of Justice, May 1, 2018. https://www.nij.gov:443/topics/courts/drug-courts/Pages/work.aspx.

Ohler, Norman. *Blitzed: Drugs in Nazi Germany.* New York: Penguin, 2017.

"Opium Throughout History." *Frontline: The Opium Kings.* Boston: WGBH, 1998. https://www.pbs.org/wgbh/pages/frontline/shows/heroin/etc/history.html.

Owen, Gary. "Standing in the Shadows: The Legacy of Harry J. Anslinger." Presented at the DEA Museum Lecture Series, October 15, 2014. https://www.deamuseum.org/wp-content/uploads/2015/08/101514-DEAMuseum-LecturesSeries-StandingintheShadows-transcript.pdf.

Padwa, Howard. *Social Poison: The Culture and Politics of Opiate Control in Britain and France, 1821–1926.* Baltimore, MD: Johns Hopkins University Press, 2012.

Page, James Lynn. "The Birth Chart of the UK—A Look at the Past and

Present." *Astro.Nu* (blog), April 12, 2017. http://www.astro.nu/2017/04/12/birth-chart-united-kingdom/.

Pagliarulo, Ned. "Gottlieb Rebuffs Pharma CEO Who Claimed Price Hikes Were 'Moral Requirement.'" Biopharmadive.com, September 12, 2018. https://www.biopharmadive.com/news/gottlieb-rebuffs-pharma-ceo-nostrum-labs-price-hikes-moral-imperative/532194/.

Pain Assessment and Management Initiative. "Pain Management and Dosing Guide." American Pain Society, n.d. http://americanpainsociety.org/uploads/education/PAMI_Pain_Mangement_and_Dosing_Guide_0228 2017.pdf.

"Papers Relating to the Foreign Relations of the United States, with the Address of the President to Congress December 2, 1913—Office of the Historian." US State Department. Accessed December 19, 2018. https://history.state.gov/historicaldocuments/frus1913/d229.

"Papers Relating to the Opium Question." Calcutta: Office of the Superintendent of Government Printing, 1870. Google Books.

Pappas, Stephanie. "Opioid Crisis Has Frightening Parallels to Drug Epidemic of Late 1800s." Live Science. Accessed December 9, 2018. https://www.livescience.com/60559-opioid-crisis-echoes-epidemic-of-1800s.html.

"Paracelsus: Alchemical Genius of the Middle Ages." *AlchemyLab.Com* (blog). Accessed January 7, 2019. https://www.alchemylab.com/paracelsus.htm.

Paracelsus and Lauron William De Laurence. *Hermetic Medicine and Hermetic Philosophy*. Chicago: De Laurence, Scott & Company, 1910. https://play.google.com/books/reader?id=CJFAAQAAIAAJ&pg=GBS.PP5.

"Paracelsus." Toxipedia. https://www.asmalldoseoftoxicology.org/paracelsus.

Paratico, Angelo. "Nero Did Not Murder Britannicus." *Beyond Thirty-Nine* (blog), July 11, 2016. https://beyondthirtynine.com/nero-did-not-murder-britannicus/.

Parkin, Tim. *Old Age in the Roman World*. Baltimore, MD: Johns Hopkins University Press, 2003.

Partlow, Joshua. "He Was America's Man in Afghanistan. Then Things Went Sour. Abdurrashi Dostum May Be Back." *Washington Post*, April 23, 2014.

Paterwic, Stephan J. *Historical Dictionary of the Shakers*. 2nd ed. Historical Dictionaries of Religions, Philosophies, and Movements. Lanham, MD: Rowman and Littlefield, 2017.

Pates, Richard, Andrew McBride, and Karin Arnold. *Injecting Illicit Drugs*. Oxford: Blackwell Publishing, 2005. Google Books.

PDMP Training and Technical Assistance Center. "Prescription Drug Monitoring Frequently Asked Questions (FAQ)." http://www.pdmpassist.org/content/prescription-drug-monitoring-frequently-asked-questions-faq.

Pearce, David. "Charles Gabriel Pravaz (1791–1853): Physician Who Developed the Hypodermic Syringe." BLTC Research. Accessed May 21, 2018. https://www.general-anaesthesia.com/people/charles-pravaz.html.

Peck, David J. "The Strange and Mysterious Death of Captain Meriwether Lewis." Lewis & Clark Medicine, 2010. http://www.lewisandclarkmedicine.com/the_strange_and_mysterious.html.

Penn, Sean. "El Chapo Speaks." *Rolling Stone*, January 10, 2016.

"The Pharmacy Museum—Universität Basel." MySwitzerland. Accessed January 7, 2018. http://www.myswitzerland.com/en/pharmazie-historisches-museum-der-universitaet-basel.html.

Phisick Medical Antiques. "Pravaz Hypodermic Syringe in Silver." Accessed May 26, 2018. http://phisick.com/item/pravaz-opiate-hypodermic-syringe-charierre/.

Pietschmann, Thomas, Melissa Tullis, and Theodore Leggett. "A Century of International Drug Control." Extended Version of World Drug Report 2008. Vienna: United Nations Office on Drugs & Crime, 2008.

Pincus, Walter. "Bush out of Loop on Iran-Contra?" *Washington Post*, September 24, 1992.

Pliny the Elder. *The Natural History Book XX: Remedies Derived from the Garden Plants*. Edited by John Bostock. Perseus Digital Library. Accessed March 26, 2018. http://www.perseus.tufts.edu/hopper/text?doc=Perseus%3Atext%3A1999.02.0137%3Abook%3D20%3Achapter%3D76#note1.

"Politics Forum." Politics Forum. Accessed April 14, 2018. https://www.politicsforum.org/.

Pollan, Michael. *Botany of Desire: A Plant's-Eye View of the World*. New York: Random House, 2001.

Popova, Maria. "The Angels and Demons of Genius: Robert Lowell on What It's Like to Be Bipolar." *Brain Pickings* (blog), February 26, 2016. https://www.brainpickings.org/2016/02/26/robert-lowell-bipolar/.

Porter, Roy, and Mikulus Teich. *Drugs and Narcotics in History*. Cambridge: Cambridge University Press, 1995. Google Books.

"The Portrait—George Washington: A National Treasure." Accessed December 30, 2017. http://georgewashington.si.edu/portrait/face.html.

Powers, Jason. "Assassins, Drug Dealers Share Fondness for Fentanyl."

Huffington Post, January 18, 2017. https://www.huffingtonpost.com/ jason-powers/assassins-drug-dealers-sh_b_9013274.html.

"Presidential Address on National Drug Policy." Washington, DC: C-Span, September 5, 1989. https://www.c-span.org/video/?8921-1/president-bush-address-national-drug-policy.

President's Commission. "On Combating Drug Addiction and the Opioid Crisis." Washington, DC: The White House, November 2017. https://www.whitehouse.gov/sites/whitehouse.gov/files/images/Final_Report_Draft_11-1-2017.pdf.

"Provisional Drug Overdose Data." Vital Statistics Rapid Release. Washington, DC: Centers for Disease Control, July 6, 2018. https://www.cdc.gov/nchs/nvss/vsrr/drug-overdose-data.htm.

"Public Enemy Number One: A Pragmatic Approach to America's Drug Problem." Richard Nixon Foundation, June 29, 2016. https://www.nixonfoundation.org/2016/06/26404/.

"publiusclodius." "Why Did King Attalus III of Pergamon Give His Country to the Roman Republic in His Final Will?" *Reddit: Ask Me Anything* (blog), 2015. https://www.reddit.com/r/AskHistorians/comments/35uzw8/why_did_king_attalus_iii_of_pergamon_give_his/.

Qian, Sima. "Records of the Grand Historian." *China History* (blog), July 22, 2017. http://history.followcn.com/2017/07/22/records-grand-historian-sima-qian/.

Queen Elizabeth I. "Elizabeth's Tilbury Speech." http://www.bl.uk/learning/timeline/item102878.html.

Quinones, Sam. *Dreamland: The True Tale of America's Opiate Epidemic.* New York: Bloomsbury Press, 2015.

Raver, Anne. "Poppies: Sowing the Seeds of a Felonious Life." *Free Lance-Star,* May 14, 1992.

Reagan, Ronald. "Radio Address to the Nation on Economic Growth and the War on Drugs," October 8, 1988. http://www.presidency.ucsb.edu/ws/index.php?pid=34997.

"Records of the Drug Enforcement Administration [DEA]." National Archives, August 15, 2016. https://www.archives.gov/research/guide-fed-records/groups/170.html.

Redford, Audrey, and Benjamin W. Powell. "Dynamics of Intervention in the War on Drugs: The Build-Up to the Harrison Act of 1914." *Independent Review,* May 15, 2016. doi.10.2139/ssrn.2533166.

Rees, Abraham. *The Cyclopaedia: Or, Universal Dictionary of Arts, Sciences, and Literature.* London: Longman, Hurst, Rees,

Orme & Browne, 1819. HathiTrust. https://catalog.hathitrust.org/Record/ 001464694?type%5B%5D=all&lookfor%5B%5D=The%20Cyclopaedia% 3A%20Or%2C%20Universal%20Dictionary%20of%20Arts%2C%20Scien ces%2C%20and%20Literature&ft=

Rein, Lisa, and Michael Horowitz. "Inner Space and Outer Space: Carl Sagan's Letters to Timothy Leary (1974)." *Timothy Leary Archives* (blog), 1974. http://www.timothylearyarchives.org/carl-sagans-letters-to-timothy-leary-1974/.

Reingruber, Agathe, Zoï Tsirtsoni, and Petranka Nedelcheva. *Going West?: The Dissemination of Neolithic Innovations between the Bosporus and the Carpathians.* London: Routledge, 2017. Google Books. https://books.google.com/books? id=gkYlDwAAQBAJ&printsec=frontcover&source=gbs_book_other_ versions_r&cad=2#v=onepage&q&f=false.

Reiss, Jonathan. "Opioid Crisis: What People Don't Know about Heroin." *Rolling Stone,* May 18, 2018.

Remington, Joseph P. *The Science and Practice of Pharmacy.* 21st ed. Edited by D. B. Troy. Philadelphia: Lippincott Williams & Williams, 2005.

Rennie, David. "Chinese Drug Addiction 'In the Genes.'" *The Telegraph,* March 15, 2001.

Reuters. "Scientists Find Gene Secret That Lets Poppies Make Morphine." *VOA,* June 25, 2015, Reuters edition. https://www.voanews.com/a/ scientists-find-gene-secret-lets-poppies-make-morphine/2837877.html.

Rice, Xan. "What Is OxyContin, the Drug behind America's Opioid Crisis?" NewStatesmanAmerica, November 19, 2017. https://www.newstatesman.com/world/2017/11/what-oxycontin-drug-behind-america-s-opioid-crisis.

Richardson, W. F. *Celsus on Medicine.* Auckland: University of Auckland, 1979. Google Books.

Ricketson, Shadrach. *Means of Preserving Health and Preventing Diseases.* New York: Collins, Perkins, 1806. Internet Archive.

Riggins, Alex. "Two Women Accused of Smuggling Arrested at Border; 81-Year-Old Allegedly Had 92 Pounds of Heroin." *San Diego Union-Tribune,* August 25, 2018, sec. News/Public Safety.

Riley, Jack. "Racial Profiling—Lessons from the Drug War." *RAND Review* 26, No. 2 (Summer 2002), 60–62. Ring, Trudy, Noelle Watson, and Paul Schellinger. *Southern Europe: International Dictionary of Historic Places.* London: Routledge, 2013. Google Books.

Robinette, G.W. *Did Lin Zexu Make Morphine.* Vol. 1. Valparaiso, Chile: Graffiti Militante Press, 2008.

Robins, Nick. *The Corporation That Changed the World: How the East India Company Shaped the Modern Multinational.* London: Pluto Press, 2012.

Robinson, Melia. "Bill Gates Once Coyly Defended LSD Use by Saying: 'I Never Missed a Day of Work.'" *Business Insider,* February 15, 2017. https://www.businessinsider.com/bill-gates-lsd-psychedelics-2017-2.

Rogers, Stan. "Was Pepper Once Worth Its Weight in Gold?" Stack Exchange [Skeptics], March 25, 2014. https://skeptics.stackexchange.com/questions/19998/was-pepper-once-worth-its-weight-in-gold.

Rosenblum, Andrew, Lisa A. Marsch, Herman Joseph, and Russell K. Portenoy. "Opioids and the Treatment of Chronic Pain: Controversies, Current Status, and Future Directions." *Experimental and Clinical Psychopharmacology* 16, no. 5 (October 2008): 405–416. doi.10.1037/a0013628.

Rothman, Lily. "Billie Holiday at 100: Her Biography Is More Complicated Than You Know." *Time,* April 7, 2015.

"Rumi Spice: A Leading Seller of Quality Afghan Saffron." RumiSpice, 2018. https://www.rumispice.com/pages/about-us.

Russo, Ethan B. "Cannabinoids in the Management of Difficult to Treat Pain." *Therapeutics and Clinical Risk Management* 4, no. 1 (February 2008): 245–259.

Sacco, Lisa N. "Drug Enforcement in the United States: History, Policy, and Trends." *Congressional Research Service,* October 2, 2014.

"San Francisco History—Population." Accessed December 17, 2018. http://www.sfgenealogy.org/sf/history/hgpop.htm.

Santacroce, Luigi, Lucrezia Bottalico, and Ioannis Alexandros Charitos. "Greek Medicine Practice at Ancient Rome: The Physician Molecularist Asclepiades." *Medicines* 4, no. 92 (December 2017). doi.10.3390/medicines4040092.

Santella, Thomas. *Opium.* Drugs—The Straight Facts. New York: Chelsea House, 2007.

Saunders, Nicholas J. *The Poppy: A History of Conflict, Loss, Remembrance & Redemption.* London: Oneworld Publications, 2013.

Scaccia, AnnaMarya. "Treating Heroin Addiction With Heroin: What You Need to Know." *Rolling Stone,* July 31, 2017.

Schadewalt, Dr. H. "Hildegard von Bingen and the Medicine of Her Time." Lecture presented at the Great Hildegard Meeting, Lake Constance, 1996. https://translate.google.com/translate?hl=en&sl=de&u=http://www.st-hildegard.com/de/st-hildegard/biografie.html&prev=search.

Schmitz, Rudolf. "Friedrich Wilhelm Sertürner and the Discovery of Morphine." *Pharmacy in History* 27, no. 2 (1985): 61–74.

Schottenhammer, Angela. "The 'China Seas' in World History: A General Outline of the Role of Chinese and East Asian Maritime Space from Its Origins to c. 1800." *Journal of Marine and Island Cultures* 1, no. 2 (December 2012): 63–86.

*Scientific American*. "Scientific American Archives." *Scientific American*. Accessed February 11, 2018. https://www.scientificamerican.com/store/archive/.

Scott, J. M. *The White Poppy*. New York: Funk & Wagnalls, 1969.

"Secret Santa." "Narco-Philanthropy." *SpectreVision* (blog), February 27, 2015. https://spectrevision.net/2015/02/27/narco-philanthropy/.

Substance Use–Disorder Prevention that Promotes Opioid Recovery and Treatment for Patients and Communities Act, H. R. 6 § (2018). https://www.finance.senate.gov/imo/media/doc/ 930%20AM%20Edits%2009.26.18%20Final%20Opioid%20Sec-by-Sec%20BIPART%20BICAM.pdf.

Serageldin, Ismail. "Ancient Alexandria and the Dawn of Medical Science." *Global Cardiology Science & Practice* 2013, no. 4 (December 30, 2012): 395–404. doi.10.5339/gcsp.2013.47.

Sexton, Anne. *The Complete Poems*. Boston: Houghton Mifflin, 1981.

*A Shaker Gardener's Manual Containing Plain Instructions for the Selection, Preparation, and Management of a Kitchen Garden, with Practical Directions for the Cultivation and Management of Some of the Most Useful Culinary Vegetables*. Boston/Cambridge: Applewood Books, 1843. HathiTrust.

Shaker Museum Mount Lebanon, "Drawer Pulls: What's Original?" December 6, 2017. https://shakerml.wordpress.com/2017/12/06/drawer-pulls-whats-original/.

Shapiro, Ari. "A Look At The Effectiveness Of Anti-Drug Ad Campaigns." *All Things Considered*. NPR, November 1, 2017. https://www.npr.org/2017/ 11/01/561427918/a-look-at-the-effectiveness-anti-drug-ad-campaigns.

Shattuck, Gary G. *Green Mountain Opium Eaters: A History of Early Addiction in Vermont*. Charleston, SC: Arcadia Publishing / The History Press, 2017.

Shields, Christopher. "Aristotle." In *The Stanford Encyclopedia of Philosophy*, edited by Edward N. Zalta, Winter ed. Stanford, CA: Metaphysics Research Lab, Stanford University, 2016.

Simpson, Sir James Y. *Anaesthesia, Hospitalism, Hermaphoroditism: And a Proposal to Stamp Out Small-Pox and Other Contagious Diseases*. Edinburgh: Adam and Charles Black, 1871. HathiTrust. https://catalog.hathitrust.org/ Record/011562519.

Singer, Dr. Jeffrey. "The Administration's Fundamental Flaw on Opioid Addiction: Talk of Progress Is Greatly Exaggerated." *New York Daily News*, October 23, 2018.

Sirin, Cigdem V. "From Nixon's War on Drugs to Obama's Drug Policies Today: Presidential Progress in Addressing Racial Injustices and Disparities." *Race, Gender & Class* 18, no. 3/4 (2011).

Sisodia, Rajeshree. "Afghanistan's Opium Babies." *Al Jazeera*, September 2, 2006. http://www.aljazeera.com/archive/2006/09/200849155726793606.html.

Skaggs, K. R. "The Form of Money: Opium in Literature and Film." *The Form of Money* (blog), February 24, 2016. http://theformofmoney-mammon.blogspot.com/2012/05/opium-series-opium-portrait-of-heavenly.html.

Smith, Van. "Baltimore's Narcotic History Dates Back to the 19th Century Shipping-Driven Boom, Quietly Aided by Bringing Turkish Opium to China." *City Paper*, October 21, 2014. http://www.citypaper.com/news/mobtownbeat/bcp-baltimores-narcotic-history-dates-back-to-the-19thcentury-shippingdriven-boom-quietly-aided-by-bring-20141021-story.html.

Smith, William French. "Organized Crime Today." *Drug Enforcement* 6, no. 1 (October 14, 1982). https://play.google.com/books/reader?id=0hNXAAAAMAAJ&printsec=frontcover&output=reader&hl=en&pg=GBS.RA4-PA7.

Solotaroff, Paul. "El Chapo: Inside the Hunt for Mexico's Most Notorious Kingpin." *Rolling Stone*, August 11, 2017.

Stafford, Lindsay. "First US-China Trade Ship Carried 30 Tons of American Ginseng." *American Botanical Council: HerbalEGram*, May 2012. http://cms.herbalgram.org/heg/volume9/05May/EmpressofChinaGinseng.html?ts=1526334496&signature=8728fee9be74129064346ed7ea37978c&ts=1526384767&signature=3b96acd42b04cc5809124ce7840db9de.

Staines, Richard. "Pfizer/Lilly Non-Opioid Pain Drug Hits Mark in Phase 3 Trial." *Pharmaphorum*, July 23, 2018. https://pharmaphorum.com/news/pfizer-lilly-tanezumab-pain-drug-hits-mark-in-phase-3-trial/.

Stanley, Theodore H. "The Fentanyl Story." *The Journal of Pain* 15, no. 12 (December 2014): 1215–1226. doi.10.1016/j.jpain.2014.08.010.

"The State of Opioid Sales on the Dark Web." *LegitScript* (blog), June 28, 2018. https://www.legitscript.com/blog/2018/06/opioid-sales-on-the-dark-web/.

Stoner, Laura E. G. "A Guide to the A. H. Robins

Company Records, 1885–2004." Richmond: Virginia Historical Society, 2008. https://www.virginiahistory.org/collections-and-resources/how-we-can-help-your-research/researcher-resources/finding-aids/ah-robins.

Stotts, Angela L., Carrie L. Dodrill, and Thomas R. Kosten. "Opioid Dependence Treatment: Options In Pharmacotherapy." *Expert Opinion on Pharmacotherapy* 10, no. 11 (August 2009): 1727–1740. doi.10.1517/14656560903037168.

Stoye, Emma. "Biotech Breakthrough as Yeast Makes Painkillers from Sugar." *Chemistry World*, August 2015.

Stroud, Ronald S. "The Gravestone of Socrates' Friend Lysis." *American School of Classical Studies at Athens* 53, no. 3 (September 1984): 355–360.

"The Struggle of Mankind Against Its Deadliest Foe." *Special Address.* New York: NBC, March 1, 1928. http://www.reefermadnessmuseum.org/otr/NarcoticEd1928.htm.

Subramanian, Lakshmi. *Parsi Traders in Western India, 1600–1900.* Oxford Research Encyclopedias. Oxford: Oxford University Press, 2018.

Subramaniyam, Lakshmi. "Transcript Expression Profiling for Adventitious Roots of Panax Ginseng, Meyer." *Gene,* 2014. doi.10.1016/j.gene.2014.05.024.

Substance Abuse and Mental Health Services Administration. "National Registry of Evidence-Based Programs and Practices (NREPP)." Accessed September 29, 2018. https://www.samhsa.gov/nrepp.

Substance Use–Disorder Prevention that Promotes Opioid Recovery and Treatment for Patients and Communities Act, Pub. L. No. H. R. 6, 250 (2018). https://www.congress.gov/115/bills/hr6/BILLS-115hr6enr.pdf.

Sulzberger, Harmann Henry (Merchant), ed. *All about Opium.* London: Wertheimer, Lea, 1884. Google Books.

Swaminathan, Nikhil. "Squabbling Over a Sumerian City." *Archaeology,* Review of Exhibition, 64, no. 2 (April 2011).

Szalavitz, Maia. "After 75 Years of AA, It's Time to Admit We Have a Problem." *Pacific Standard,* February 10, 2014. https://psmag.com/social-justice/75-years-alcoholics-anonymous-time-admit-problem-74268.

———. "One Hundred Years Ago, Prohibition Began in Earnest—and We're Still Paying for It." *Pacific Standard,* January 2, 2015. https://psmag.com/social-justice/one-hundred-years-ago-prohibition-began-earnest-still-paying-97243.

T, Buddy. "The Various Types of Heroin: Highly Addictive Heroin Has Made a Comeback in the United States." *Verywell Mind,* July 8, 2018. https://www.verywellmind.com/heroin-photos-4020361.

Tax Foundation. "Cigarette and Tobacco Taxes." *Tax Foundation* (blog). Accessed June 30, 2018. https://taxfoundation.org/individual-consumption-taxes/excise-taxes/cigarette-and-tobacco-taxes/.

Taylor, Bryan. "The First and the Greatest: The Rise and Fall of the Vereenigde Oost-Indische Compagnie." *Global Financial Data*, November 6, 2013. https://www.globalfinancialdata.com/GFD/Blog/1st-and-greatest-rise-and-fall-vereenegde-oost-indische-co.

Taylor, David A. "Getting to the Root of Ginseng." *Smithsonian*, July 2002.

Terry, Charles Edward, Mildred Pellens, and Committee on Drug Addiction. *The Opium Problem*. Vol. 16. New York: Bureau of Social Hygiene (New York), 1928. HathiTrust. https://catalog.hathitrust.org/Record/001133723?type%5B%5D=all&lookfor%5B%5D=The%20Opium%20Problem&ft=.

Terry, Mark. "In Negotiations of 1,000 Opioid Lawsuits, Purdue Pharma Offers Free Opioid Therapy." BioSpace, September 12, 2018. https://www.biospace.com/article/-jc1n-in-negotiations-of-1-000-opioid-lawsuits-purdue-pharma-offers-free-opioid-therapy/.

Thacher, James. *The American New Dispensatory*. Boston: T.R. Wait, 1810. http://catalog.hathitrust/079/record/008890890.

Theophrastus. *Enquiry into Plants and Minor Works on Odours and Weather Signs*. Translated by Arthur Hort. London: W. Heinemann, 1916. Internet Archive.

"Thirteen Celebrated Geniuses and Their Shocking Drugs of Choice." *BrainJet*, April 10, 2015. https://www.brainjet.com/random/2357520/13-celebrated-geniuses-and-their-shocking-drugs-of-choice/.

"Thomas Jefferson's Monticello." Thomas Jefferson's Monticello. Accessed December 29, 2017. http://home.monticello.org/.

Thorpe, Vanessa. "How Did 18th Century's Literary Women Relive Domestic Distress? With Opiates." *The Guardian*, March 11, 2018, The Observer edition, sec. Books.

Tibi, Selma. "Al-Razi and Islamic Medicine in the 9th Century." *Journal of the Royal Society of Medicine* 99, no. 4 (April 2006): 206–207.

"Timeline of Selected FDA Activities and Significant Events Addressing Opioid Misuse and Abuse." Washington, DC: US Food and Drug Administration, August 2018. https://www.fda.gov/downloads/Drugs/DrugSafety/InformationbyDrugClass/UCM566985.pdf.

Tiwari, Pawan. "Grocer Sells Ayurveda Medicines with Opium, Held." *Times of India*, April 30, 2017, sec. City.

"TOXNET." Accessed January 20, 2019. https://toxnet.nlm.nih.gov/cgi-bin/sis/search2/r?dbs+hsdb:@term+@rn+@rel+437-38-7.

Treaster, Joseph B. "Four Years of Bush's Drug War: New Funds but an Old Strategy." *New York Times*, July 28, 1992, sec. U.S.

Trickey, Erick. "Inside the Story of America's 19th-Century Opiate Addiction." *Smithsonian*, January 4, 2018.

Turner, Jack. "The Spice That Built Venice." *Smithsonian*, November 2, 2015.

University of Southern California. "A Nonaddictive Opioid Painkiller with No Side Effects." *Cell & Microbiology*, January 5, 2018. https://phys.org/news/2018-01-nonaddictive-opioid-painkiller-side-effects.html.

Unschuld, Paul. *Medicine in China: A History of Pharmaceuticals*. Berkeley: University of California Press, 1986. Google Books.

"Urges Taxpayers to Aid Government." *New York Times*, October 6, 1917, Special to The New York Times edition, sec. Archives. https://www.nytimes.com/1917/10/06/archives/urges-taxpayers-to-aid-government-commissioner-roper-appeals-to.html.

US Department of Health and Human Services. "What Is the U.S. Opioid Epidemic?" Health and Human Services. Accessed August 18, 2018. https://www.hhs.gov/opioids/about-the-epidemic/index.html.

"Use of Opium and Cannabis in the Traditional Systems of Medicine in India." New York: United Nations Office on Drugs & Crime, 1965. https://www.unodc.org/unodc/en/data-and-analysis/bulletin/bulletin_1965-01-01_1_page004.html.

Van Dyke, Paul A. *The Canton Trade*. Hong Kong: Hong Kong University Press, 2005.

Van Zee, Art. "The Promotion and Marketing of OxyContin: Commercial Triumph, Public Health Tragedy." *American Journal of Public Health* 99, no. 2 (February 2009): 221–227. doi.10.2105/AJPH.2007.131714.

Vanderbilt University Medical Center. "Neonatal Abstinence Syndrome and Opioid Policy." Accessed January 21, 2019. https://www.vumc.org/nas/.

Veiga, Paula. "Opium: Was It Used As a Recreational Drug in Ancient Egypt?" PhD thesis, Institut für Ägyptologie, Ludwig-Maximilians-Universität, 2017. https://www.openstarts.units.it/bitstream/10077/14300/1/ATRA_3_online-14_Veiga.pdf.

Veith, Ilza. *The Yellow Emperor's Classic of Internal Medicine*. Foreword copyright Linda I. Barnes. Oakland: University of California Press, 2016. Originally published 1949 by Williams & Wilkins.

Vendel, Ottar. "The Spirits of Nature: Religion of the Egyptians." Absolute Egyptology. Accessed January 7, 2018. http://www.nemo.nu/ibisportal/0egyptintro/1egypt/index.htm.

Verthein, Uwe, Karin Bonorden-Kleij, Peter Degkwitz, Christoph Dilg, Wilfried K. Köhler, Torsten Passie, Michael Soyka, Sabine Tanger, Mario

Vogel, and Christian Haasen. "Long-Term Effects of Heroin-Assisted Treatment in Germany." *Addiction* 103, no. 6 (June 1, 2008): 960–966. doi.10.1111/j.1360-0443.2008.02185.x.

Villa, Lauren M. P. H. "Methadone and Suboxone: What's the Difference Anyway?" *DrugAbuse*, December 27, 2016. https://drugabuse.com/suboxone-vs-methadone/.

Vines, Stephen. "The Decline of Compradors." *South China Morning Post*, August 20, 2012, International edition. https://www.scmp.com/business/money/article/1018386/decline-compradors.

Waley, Arthur. *The Opium War Through Chinese Eyes*. Stanford, CA: Stanford University Press, 1958.

Ward, Geoffrey. *Before the Trumpet*. New York: Random House, 1985.

Ward, Geoffrey C., and Ken Burns. *The Vietnam War: An Intimate History*. New York: Alfred A. Knopf, 2017.

Warner, Mallory. "The Power of the Poppy: Exploring Opium through 'The Wizard of Oz.'" National Museum of American History, November 9, 2016. http://americanhistory.si.edu/blog/opium-through-wizard-oz.

Wasacz, John. "Natural and Synthetic Narcotic Drugs." *American Scientist* 69, no. 3 (June 1981): 318–324.

Wasson, L. Gordon, Albert Hofmann, and Carl A. P. Ruck. *The Road to Eleusis*. Twentieth Anniversary Edition. Los Angeles: William Dailey Rare Books, 1998.

Wei, Frances Loke. "Chinese Pidgin English: You Sabee?" Unravel, December 22, 2017. https://unravellingmag.com/articles/chinese-pidgin-english/.

Weil, Andrew, and Winifred Rosen. *From Chocolate to Morphine*. Boston: Houghton Mifflin, 1993.

Wen, Philip. "China Says U.S. Should Do More to Cut Its 'Enormous' Opioid Demand." *Reuters*, December 29, 2017.

Werb, D., A. Kamarulzaman, M. C. Meacham, C. Rafful, B. Fischer, S. A. Strathdee, and E. Wood. "The Effectiveness of Compulsory Drug Treatment: A Systematic Review." *International Journal of Drug Policy* 28 (February 1, 2016): 1–9. doi.10.1016/j.drugpo.2015.12.005.

"What's This Agonist /Antagonist Stuff?" National Alliance of Advocates for Buprenorphine Treatment. Accessed June 13, 2018. https://www.naabt.org/faq_answers.cfm?ID=5.

Whigham, Nick. "Silk Road Founder Ross Ulbricht Sends Messages from Prison." *News Pty Limited*, 2018. https://www.news.com.au/technology/online/social/contact-with-the-outside-is-good-for-me-silk-road-founders-messages-to-the-world/news-story/fea647f5ec53172c8cb47db81ee29971.

White, A. J. *Shaker Almanac, 1884.* New York: A. J. White, 1883. HathiTrust. https://catalog.hathitrust.org/Record/100184558?type %5B%5D=all&lookfor%5B%5D=Shaker%20Almanac%2C%201884&ft=.

White, C. Michael. "Pharmacologic and Clinical Assessment of Kratom." *American Journal of Health-System Pharmacy* 75, no. 5 (March 2008): 261–267. doi.10.2146/ajhp161035.

"Why Imprisonment Is More Harm than Help to Addicted Offenders." Skywood Recovery, April 23, 2018. https://skywoodrecovery.com/why-imprisonment-is-more-harm-than-help-to-addicted-offenders/.

Wigal, Donald. *Opium: The Flowers of Evil.* New York: Parkston Press, 2014.

Wikipedia Contributors. "Averroes." Accessed October 29, 2018. https://fr.wikipedia.org/w/index.php?title=Averro%C3%A8s&oldid=146530971.

———. "Battle of Chuenpi." Accessed March 12, 2018. https://en.wikipedia.org/w/index.php?title=Battle_of_Chuenpi&oldid=805002069.

———. "The Thirteen Factories." Accessed November 29, 2017. https://en.wikipedia.org/w/index.php?title=Thirteen_Factories&oldid=812795255.

———. "Zhang Qian." Accessed March 17, 2018. https://en.wikipedia.org/w/index.php?title=Zhang_Qian&oldid=830942783.

Wilkins, John S. "What's in a Name?" *The Renaissance Mathematicus* (blog), May 1, 2012. https://thonyc.wordpress.com/2012/05/01/whats-in-a-name-2/.

Willasey-Wilsey, Tim. "The Enigmatic Warren Hastings and His Calcutta Properties." The Victorian Web, 2014. http://www.victorianweb.org/history/empire/india/74.html.

Williams, Senator John J. "Manufacture of Smoking Opium." Washington, DC: US Senate, November 22, 1913. https://www.finance.senate.gov/imo/media/doc/SRpt63-130.pdf.

Wilson, Gaye. "Medicine (Thomas Jefferson Encyclopedia)." Thomas Jefferson Foundation, February 2004. https://www.monticello.org/site/research-and-collections/medicine.

Wisniewski, J. "Five Myths about the Revolutionary War Everyone Believes." *Cracked* (blog), March 16, 2013. http://www.cracked.com/article_20306_5-myths-about-revolutionary-war-everyone-believes.html.

Withington, Edward Theodore. *Medical History from the Earliest Times: A Popular History of the Healing Art.* London: Scientific Press, 1894.

"The Wizard of Oz: Five Alternative Readings." BBC, August 19, 2014. http://www.bbc.com/culture/story/20140819-the-wizard-of-oz-hidden-meanings.

Wong, Szu Schen. "Syphilis and the Use of Mercury." *Pharmaceutical Journal* (blog), September 8, 2016.

"World Drug Report 2008." United Nations Office on Drugs & Crime, 2008. https://www.unodc.org/unodc/en/data-and-analysis/WDR-2008.html.

"The World Factbook: 2013–2014." Central Intelligence Agency, 2013. https://www.cia.gov/Library/publications/the-world-factbook/fields/2086.html.

Wright, C. R. A., and G. H. Beckett. "On the Action of Organic Acids and Their Anhydrides on the Natural Alkaloids. Part I." *Journal of the Chemical Society* 4 (1875): 1031–1043.

Wright, Hamilton, Lloyd Bryce, and Gerrit John Kollen. "Report to the Secretary of State by the American Delegates." Presented at the Second International Opium Conference, The Hague, December 2, 1913. https://history.state.gov/historicaldocuments/frus1913/ch30.

Wright, Katherine. "Where Did Syphilis Come From?" *The Guardian*, October 26, 2013.

Wright, Thomas, ed. *The Travels of Marco Polo The Venetian.* Translated by William Marsden. London: H.G. Bohn, 1854. HathiTrust. https://catalog.hathitrust.org/Record/007670407?type%5B%5D=all& lookfor%5B%5D=Marco%20Polo%20the%20Venetian&ft=.

Xin, Hao. "What Are Some Mind-Blowing Facts and Stories about China's Emperors in History?" Quora, August 9, 2017. https://www.quora.com/What-are-some-mind-blowing-facts-and-stories-about-Chinas-Emperors-in-history#YjtHC.

Yates, Matthew Tyson, ed. "General Conference of the Protestant Missionaries of China." Shanghai: Presbyterian Mission Press, 1877. https://archive.org/details/recordsofgeneral00gene/page/n7.

Yeh, Brian T. "Drug Offenses: Maximum Fines and Terms of Imprisonment for Violation of the Federal Controlled Substances Act and Related Laws." *Congressional Research Service*, January 20, 2015, 17.

Zexu, Lin. "Letter of Advice to Queen Victoria." In Teng and Fairbank, *China's Response to the West*. Cambridge, MA: Harvard University Press, 1954.

Zheng, Yangwen. "The Social Life of Opium in China, 1483–1999." *Modern Asian Studies* 37, no. 1 (2003): 1–39.

# Index